Hotel & Restaurant Design No. 2

Hotel & Restaurant Design NO.2

国际酒店与餐厅经典设计 2

[美] 罗杰·易　编著

中国建筑工业出版社

著作权合同登记图字：01-2008-6178号

图书在版编目（CIP）数据

国际酒店与餐厅经典设计 2/（美）易编著. —北京：中国建筑工业出版社，2009
ISBN 978-7-112-10684-4

Ⅰ.国... Ⅱ.易... Ⅲ.①饭店－建筑设计－世界－图集 ②餐厅－建筑建设－世界－图集 Ⅳ.TU247-64

中国版本图书馆CIP数据核字（2009）第021704号

责任编辑：程素荣

Hotel & Restaurant Design No.2
国际酒店与餐厅经典设计2
[美] 罗杰 · 易 编著
*
中国建筑工业出版社出版、发行（北京西郊百万庄）
各地新华书店、建筑书店经销
北京嘉泰利德公司制版
北京盛通印刷股份有限公司印刷
*
开本：965×1270毫米 1/16 印张：15¾ 字数：550千字
2009年5月第一版 2009年5月第一次印刷
定价：**258.00**元
ISBN 978-7-112-10684-4
(17617)

Contents

Introduction

One Moment in Paradise

Why do new hotel and restaurant environments seem so much better than reality that we can't have enough of them?

Is America ready to see the beloved image painted by Norman Rockwell, *Freedom from Want* (1943), with its festive, multi-generational family seated at the Thanksgiving table watching Grandma and Grandpa set down the roast turkey, take place in an Appleby's or Four Seasons? Grandparents of the World War II generation wouldn't hear of it. If the hospitality industry has its way, however, Grandma, Grandpa and everyone else will soon be donning their coats and heading out the door. The latest designs of hotels, restaurants and other outposts of the hospitality industry are making the switch seductively easy.

Right now, our sentimentality keeps us seated. But there are obvious reasons why we're spending more time in hotels and restaurants. Business travelers are on the road more, for example, keeping an expanding global economy connected. Statistics from Smith Travel Research show that the business sector's demand for hotel rooms reversed steep declines after 9/11 by rising 3.4 percent in 2004, 2.6 percent in 2005 and 0.8 percent in 2006. Busy individuals and two-income families also find themselves away from home so much that a hefty 47.9 percent of each food dollar is now spent in restaurants and other food services.

Yet there are numerous reasons why we want to spend more time in hotels and restaurants even when we don't need to. First, as the developed world's top workaholics, many Americans feel overworked. The European Community's Office of Economic Cooperation and Development notes that we logged 1,825 hours on the job in 2004, outdoing the Japanese (1,789 hours), the British (1,669 hours), the Germans (1,443 hours) and the French (1,441 hours).

All this eager beavering leaves one in five adults struggling with daytime sleepiness, and 50 percent of those aged 18 to 34 admitting that daytime sleepiness actually interferes with their work. Adding to our discomfort, more than half of all employees experience high levels of stress, 50 percent report that they miss one or two days of work each year because of stress, and 46 percent complain they arrive at work one to four days a year too stressed to be effective.

How is the hospitality industry responding to all this? By letting us immerse ourselves, however briefly, in an increasingly idealized world where everything runs smoothly, people are eager to please, the food is fabulous, cleaning up is somebody else's chore, the architecture is impressive, and the interior design is stylish, efficient and comfortable. That's no accident, thanks to the industry's current emphasis on brand building and unique guest experiences, which is producing a bumper crop of environments we find hard to forget or leave. Just imagine how happy you'd be arriving at the splendid new hotels and restaurants, designed by leading U.S. architects and interior designers, that appear in the following pages of *Hotel and Restaurant Design No. 2*. Millions of others are doing the same—and then heading for the door. Grandma and Grandpa, you're going too?

Roger Yee
Editor

Aria Group Architects, Inc.

830 North Blvd.
Oak Park, IL 60301
1.708.445.8400
1.708.445.1788 (Fax)
www.ariainc.com

Aria Group Architects, Inc.

California Pizza Kitchen Foxwoods
Foxwoods Resort Casino
Mashantucket, Connecticut

Want to relax with friends over sizzling, Jamaican Jerk-style pizza, served with great side orders and beverages, between exciting gaming sessions? A leader in California-style cuisine since 1985, California Pizza Kitchen has brought its signature pizzas, pastas, soups, sandwiches, appetizers and desserts to Foxwoods Resort Casino, in Mashantucket, Connecticut. To give the new 166-seat, 5,270-square-foot restaurant visibility in an upper floor concourse where everything is routinely grandiose, California Pizza Kitchen Foxwoods has been designed by Aria Group Architects to resemble a freestanding building whose expansive windows, stone piers, and metal canopy welcome guests from afar. The open dining room draws energy from the exhibition kitchen that serves as the focal point for concentric rings of counter, booth, banquette and table seating. Like the innovative menu, the design offers thoughtful detailing to enhance the dining experience. A trellis over the exhibition kitchen, for example, dramatizes the food preparation. Pendant lighting fixtures, concealed uplighting and recessed downlights help set the wall and ceiling planes in motion. Materials and furnishings such as fieldstone, wood, limestone tile, and wood and upholstered furniture impart a casual, contemporary feel. Sharing casino patrons' enthusiasm, California Pizza Kitchen proudly displays the new restaurant in its corporate communications.

Top: Exterior.

Left: Main dining room and exhibition kitchen.

Bottom: Bar.

Opposite top left: Rear of main dining room.

Opposite top right: View of interior from entrance.

Photography: AG Photography.

Aria Group Architects, Inc.

Hyatt Regency McCormick Place
Forno Café, Shor Restaurant, M/X Lounge, Board Room
Chicago, Illinois

Right: Shor Restaurant.

Below right: Forno Café.

Bottom right: Board Room.

Opposite: M/X Lounge.

Photography: Ballogg Photography.

Convention goers, other business travelers, and vacationers willing to trade off-season deals for proximity to downtown Chicago are happily booking into the 800-room Hyatt Regency McCormick Place, adjacent to the Windy City's mammoth McCormick Place convention center. While the Loop is just minutes away by car or cab, guests seeking hotel food and beverage services needn't compromise, thanks to the opening of four new venues, designed by Aria Group Architects. In replacing an existing buffet, these facilities provide more flexibility for space utilization, as well as wider variety for guests. Separate spaces are differentiated by both décor and degree of privacy, beginning with exposure to a major corridor linking the Hyatt Regency and McCormick Place, and proceeding through a series of "screened" layers that become progressively more opaque. Thus, Forno Café presents itself as a fashionable, minimally detailed Italian trattoria ideal for quick service, Shor Restaurant is a Chicago grill where trellises and planting boxes create a breezy, outdoor atmosphere, M/X Lounge gives guests a chic modern setting with flowing space and smart-looking furnishings, and Board Room establishes an intimate and immaculately tailored setting for one-on-one and small gatherings. Whatever their whims, guests will find this foursome eager to please.

Aria Group Architects, Inc.

Wildfire
Perimeter Mall
Atlanta, Georgia

Right: Ellington Room.

Below left: Outdoor café.

Below right: Bar.

Opposite top left: Exterior at main entrance.

Opposite top right: Outdoor dining patio.

Photography: Ballogg Photography.

With millions of TV viewers following *The War*, Ken Burns' gripping documentary of World War II, America's fascination with the 1940s continues unabated. That's why Atlantans are enjoying steaks, chops and seafood in a vivacious setting redolent of a 1940s dinner club—where Duke Ellington, Tommy Dorsey or the Andrews Sisters might appear at any moment—now that Wildfire, a 416-seat, 11,518-square-foot restaurant, has opened in their city's Perimeter Mall. Designed by Aria Group Architects for Lettuce Entertain You Enterprises, the lovingly detailed restaurant accommodates an entry lobby, bar, dining rooms, private party rooms and outdoor dining within and around a new, freestanding building, combining the vision of Lettuce's noted restaurateur Rich Melman with Aria Group's exuberant interpretation of 1940s style. Its bright, contemporary spaces are characterized by clean lines, a color palette of blond woods, chocolates and greens, millwork and accents in quarter-sawn oak, natural stone and cultured stone, transitional furniture, closely coordinated carpet, draperies, upholstery and wallcoverings, terrazzo floors, and custom decorative light fixtures from New Metal Crafts. Add vintage jazz recordings to this memorable scene, and guests have the perfect accompaniment for such up-to-date Wildfire entrées as filet medallion trio and Macadamia nut crusted halibut.

WILDFIRE
Steaks Chops Seafood
94

Aria Group Architects, Inc.

Republic Pan Asian Restaurant and Lounge
Chicago, Illinois

Above: Stairway Foyer.

Left: First floor bar lounge and dining area.

Below right: Booth seating in second floor dining room.

Bottom right: Second floor banquette seating and community table.

Photography: Ballogg Photography.

Jeff Zhang and Sandy Yu, husband-and-wife restaurateurs from Shanghai whose previous Chicago ventures established a reputation for delectable Asian fare and impeccable service, have taken the boldest step yet in their culinary career by opening Republic, a 342-seat, two-level, 10,000-square-foot Pan Asian restaurant and lounge, designed by Aria Group Architects, in the trendy River North neighborhood. The first-floor bar lounge and dining area and second floor dining room, private dining room and sake lounge are open, comfortable and unabashedly hip, keeping pace with creative entrées drawn from Korean, Thai, Vietnamese, Japanese, Chinese and Polynesian tradition. What gives Republic its aesthetic panache is its ability to extend itself vertically through strategically placed ceiling coves that add depth and complexity, expand its horizontal limits with wall niches, light washed walls and sweeping views from full-height windows, and project a stylish yet minimal environment, playing curved forms against orthogonal spaces with such basic elements as wood, textured wallcoverings, terrazzo, vinyl "wood" floors, classic modern furniture, bold and colorful fabrics, recessed and concealed lighting, drywall and paint. All this and, as the *Chicago Tribune* raves, great food at reasonable prices, have made Republic as irresistible as its Pad Thai, Kobe shoulder loin, and satay.

BBG-BBGM

515 Madison Avenue
4th Floor
New York, NY 10022
212.888.7663
212.935.3868 (Fax)
www.bbg-bbgm.com

1825 K Street NW
Suite 300
Washington, DC 20006
202.452.1644
202.452.1647 (Fax)

20645 North Pima Road
Suite 205
Scottsdale, AZ 85255
480.538.3288
480.538.3289 (Fax)

124 Jersey Road
Paddington NSW 2021
Australia
612.9362.4055
612.9362.4955 (Fax)

1266 Nanjing Road West
Plaza 66, 39/F
Jing An District
Shanghai 200040 China
86.21.6103.8428
86.21.6103.8565 (Fax)

BBG-BBGM

The Lodge at Turning Stone
Turning Stone Resort & Casino
Verona, New York

You needn't be a harried New Yorker to appreciate the pleasures of a new, five-star boutique hotel, The Lodge at the Turning Stone Resort & Casino, located four hours north of the Big Apple in Verona, New York. Designed by BBG-BBGM to evoke the timber-and-stone architecture of the Adirondacks on four floors totaling 104,000 square feet, the award-winning Lodge offers 98 luxury suites, including a 2,900-square-foot Presidential Suite, in a warm, intimate setting that features a lobby, Great Room with 26-foot ceiling and stone fireplace, and AAA 4-Diamond Wildflowers restaurant and bar. Interestingly, The Lodge complements Turning Stone, New York State's first casino (1993), while maintaining its own identity on the 1,200-acre resort's landscaped grounds. It's part of the Oneida Indians' strategy to develop businesses that are independently successful, including the existing Casino and Tower, 800-seat theater, numerous food services, championship golf course, and the newest addition, Skana: The Spa at Turning Stone. Thus, The Lodge's interiors create a friendly, open atmosphere that is almost turn-of-the-century in feeling, using cherry wood, fieldstone, transitional furnishings and an array of lighting fixtures. Getting away from it all or making the scene, guests find The Lodge can accommodate any itinerary.

Top left: Main entry.

Top right: Bar.

Upper right: Presidential Suite living room.

Above right: Presidential Suite master bedroom.

Right: Concierge.

Opposite: Great Room.

Photography: Mike Butler.

BBG-BBGM

InterContinental Boston
Boston, Massachusetts

Top left: Lobby and Library.

Above far left: Reception desk.

Above left: Spa lounge.

Below far left: Guestroom.

Below left: Presidential Suite bath.

Above: Champagne Bar at RumBa.

Right: Private dining room of Miel, Brasserie Provençal a French restaurant.

Photography: Mike Butler.

Nothing could be more agreeable for Bostonians and their guests than to relax among book-lined walls and a fireplace in The Library, which faces the reception desk in the lobby of the new, 424-room InterContinental Boston. Located along the Four-Point Channel Canal near the Financial District, the InterContinental Boston clearly knows the needs of business and leisure visitors to this storied community. The conference center, restaurant, bars, spa, 386 guestrooms and 38 suites, occupying floors 1-12 of a building topped by luxury residences, have been designed by BBGM to blend historic elegance and contemporary style, giving guests a rich variety of grand spaces with fine detailing. The soaring lobby, for example, makes guests feel welcome with conveniently placed main and service desks of mahogany and marble, clear glass display cabinets acting as room dividers, and a lounge and lobby bar where the floor is raised and the ceiling is dropped. Guestrooms, graciously appointed in mahogany millwork and casegoods, textiles, carpet and artwork, incorporate a distinctive design feature, windows placed above the bed, to enhance the guest experience by bringing daylight into the bathroom.

BBG-BBGM

Swissôtel Krasnye Holmy
Moscow, Russia

Muscovites can proudly identify one of their city's newest landmarks, the five-star Swissôtel Krasnye Holmy, even if they never visit the 34-story, 235-room structure. It's not just that the hotel's sleek metal and glass cylinder reshapes a skyline dominated by Stalin's seven towers. The sophisticated interiors, designed by BBGM, bring modern, world-class luxury accommodations to a city that still provides relatively few to business travelers. Incorporating a business center and 11 meeting rooms for business conferences and social events, Swiss Executive Club Lounge, two restaurants, two bars, spa and wellness center, in addition to a reception area, lobby/lounge and range of plush guestrooms and suites, the interiors embrace their curvilinear geometry and soaring views to create an unforgettable experience for guests. Executive facilities on the 29th floor, for example, accommodate everything from an intimate, 10-seat boardroom to a 112-seat meeting room in a setting that combines flexible dimensions, stylish contemporary furnishings and advanced communications with breathtaking views of the Kremlin and Red Square. Not to be outdone, even the most modest guestrooms offer generous dimensions, luxurious furnishings and showcase baths along with spectacular views. Executives who seek five-stars lodgings in Moscow can now add the Swissôtel Krasnye Holmy to their lists.

Top left: Presidential suite master bedroom.

Top right: Concerto Ristorante.

Left: City Space Bar.

Below left: Zurich Ballroom.

Bottom left: Presidential Suite bath.

Opposite: Lobby.

Photography: Adam Parker.

BBG-BBGM

Westin Boston Waterfront
Boston, Massachusetts

Below left: Mezzanine of lobby.

Right: Sauciety Restaurant.

Below middle right: Prefunction area.

Bottom right: Birch Bar.

Photography: Nikolas Koeing.

Even the most hurried, nononsense business traveler pauses to gaze at the "New England Garden" that defines the lobby of the new, 793-room Westin Boston Waterfront with artist-crafted birch trees. It's only natural. While the fresh, contemporary interiors, designed by BBGM, cater to guests' needs with 32,000 square feet of meeting space, Sauciety Restaurant, Birch Bar, and fitness center, all directly connected to the new, 516,000-square-foot Boston Convention & Exhibition Center and conveniently close to Logan Airport, the design recognizes there is more to life than business. The "Garden" theme, employing natural materials, comfortable furnishings, and naturalistic lighting, instills the hotel with structure and a sense of place to remind guests of the beauty and heritage of the New England landscape surrounding them. The lobby, for example, is a majestic space where trees, lounge seating, and a sweeping, wood-paneled soffit suggest an open field. The Birch Bar is dressed in an aquaglass panel and accented with a wood fence as a façade. The 150-seat Sauciety Restaurant features divider walls of woven leather and metal, and backlit glass feature wall units displaying hand-blown glass and natural wood objects. Yes, the Westin Boston Waterfront is all about business—and more.

BraytonHughes Design Studios

639 Howard Street
San Francisco, CA 94105-3926
415.291.8100
415.434.8145 (Fax)
www.bhdstudios.com
info@bhdstudios.com

BraytonHughes Design Studios

Mirabel
North Scottsdale, Arizona

At 3,000 feet above sea level, the views of Phoenix and the Valley of the Sun are incredible at Mirabel, a 713-acre, member-owned country club and residential community in North Scottsdale, Arizona. But the panorama is only one reason prospective residents are drawn to this new development by Discovery Land Company, which is building a Tom Fazio-designed golf course and such club amenities as a desert lodge-style clubhouse, spa facilities, tennis courts, and fitness center to accompany 309 custom home sites and 38 golf villas. The recently completed, 35,000-square-foot clubhouse, featuring interior architecture and interior design by BraytonHughes Design Studios, plays a pivotal role in inaugurating the project. Its numerous features, including a formal entry and lobby, salon, mixed grill, private dining, outdoor dining, fitness center, pool, men's and women's lounges and lockers, and pro shop, project an irresistible image of comfort, charm, and quality, all accented with the flavor of the Arizona landscape. From its stone piers and fireplaces, ceramic tile floors, and Mesquite wood millwork to its Axminster carpet, desert rustic furnishings, and wrought-iron-and-glass light fixtures, the clubhouse welcomes the 300 club members and their families and friends to the good life that awaits them.

Top right: Exterior.

Above: Outdoor fireplace lounge.

Left: Club room.

Opposite: Mixed grill with view of bar.

Architect: Hart Howerton.

Photography: Anthony Gomez.

BraytonHughes Design Studios

Market Bar
San Francisco, California

San Franciscans are serious foodies, so top-rated Bay Area restaurants typically favor interiors that celebrate good design without overpowering their cuisine and service. This is certainly true of the new, 139-seat, 3,162-square-foot Market Bar, part of the renovation and transformation of the historic San Francisco Ferry Building into a popular dining venue and fresh food market. What restaurateurs Doug Biederbeck and Joseph Graham wanted from BraytonHughes Design Studios was a non-themed restaurant that could be seen as a contemporary of the original structure. The design meets this goal so convincingly that guests may never realize how new or compact Market Bar really is. The timeless imagery greeting guests is a classic San Francisco dining room, finely tailored in walnut wood, stone tile, glazed ceramic tile, pressed tin ceiling, classic wooden bistro furniture, and pendant lighting fixtures. To accommodate a bar, two symmetrically placed dining rooms, outdoor dining, kitchen, storage and office, the design team added an unobtrusive mezzanine. Customers enjoy Mediterranean-style entrees such as grilled marinated swordfish with Helda beans, pancetta and shallots, and tomato beurre blanc in a setting where legendary newspaper columnist Herb Caen would have instantly felt at home.

Top left: Banquette seating.

Top right: Outdoor dining flanking Ferry Building.

Above: Detail of banquette.

Right: Dining room with Ferry Building arcade.

Opposite: Bar.

Photography: John Sutton.

BraytonHughes Design Studios

Vaquero
Westlake, Texas

Imagine life in a private, secure, and exclusive family community with a world-class Tom Fazio-designed golf course and traditional equity country club that includes a large swim center, fishing ponds, spa and fitness center, kids' camp, tennis courts, and an elegant clubhouse at its center, encircled by luxury homes. For 385 members of Vaquero and their families, this ideal is becoming a reality on 525 acres of the former Circle T Ranch in Westlake, Texas, minutes from Dallas, Fort Worth and DFW International Airport. Naturally, the 34,000-square-foot clubhouse sets the tone for the community, a development of the Discovery Land Company. Its interior architecture and interior design, created by BraytonHughes Design Studios, encompasses an entry lobby, mixed grill, outdoor dining, spa, men's and women's lounges and lockers, and pro shop that constitute a showcase for the contemporary Texas ranch manor lifestyle. Keeping advanced technology accessible but concealed, the design celebrates local history and tradition with grand central spaces and intimate smaller areas appointed in alder wood paneling, plaster walls, reclaimed oak floors, limestone entry floor, ranch-style furnishings, iron and colored glass accents, and artisan lighting fixtures.

Top left: Corridor.

Top right: Pro shop.

Above right: Lounge.

Bottom right: Bar in mixed grill.

Opposite: Fireplace in entry lobby.

Architect: Hart Howerton.

Photography: John Sutton.

C. David Robinson Architects
BraytonHughes Design Studios

Cliff House
San Francisco, California

Whether they come for the seafood or the view, the residents of San Francisco have been as eager as the tourists to visit the Cliff House's dining rooms since 1863. Now, in another chapter of the fabled establishment's history, the fourth version of the Cliff House has added a new, 320-seat, two-story dining room/ballroom as part of a 26,000-square-foot historic renovation and expansion, designed by C. David Robinson as architect and interior architect and BraytonHughes Design Studios as interior designer. Using a nautical theme to tie the new contemporary space to an historic existing café, the design is configured to maximize the Cliff House's fabled view of the Pacific Ocean. For this reason, the bar and serving stations are located on an upper tier at the back, while guests on the main dining floor enjoy unobstructed views through the addition's floor-to-ceiling glass walls. Careful attention was given to materials and furnishings to complement the views, and it shows. From the bleached oak paneling, hand-cast glass bar top, and mahogany floor to the contemporary furnishings and custom-made chandeliers, the Cliff House has created yet another reason for San Franciscans to return again and again, with or without eager out-of-towners in tow.

Top left: Dining room seen from upper story.

Top right: Pacific Ocean view from dining room.

Left: Bar.

Photography: Richard Barnes.

Cleo Design

3571 Red Rock Street
Suite C
Las Vegas, NV 89103
702.873.7070
702.220.4838 (Fax)
www.cleo-design.com

Cleo Design

Beau Rivage Resort & Casino
Jia Restaurant
Biloxi, Mississippi

Welcome back! This joyous greeting for guests of the 1,740-room Beau Rivage Resort & Casino, in Hurricane Katrina-ravaged Biloxi, Mississippi, concludes a top-to-bottom, renovation by celebrating such highlights as three stunning new gourmet restaurants. One of the restaurants, Jia, capitalizes on America's growing appetite for Asian cuisine by offering a delectable variety of regional specialties from Thailand, Japan, China and Vietnam. Its dynamic, 4,252-square-foot interior, designed by Cleo Design, offers guests four distinct dining areas, including an 82-seat dining room, 12-seat lounge/bar, 54-seat teppanyaki and 18-seat sushi bar that shape the guest experience by providing a unique setting for each function without detracting from the restaurant's overall unity. The secret ingredient: This timely, Asian-inspired space carefully coordinates the bold use of such natural materials as bamboo and teak, contemporary, crisply-tailored furnishings, custom-designed lighting fixtures, a red waterfall at the entrance, and art features in all four areas to interpret a shared aesthetic vision. In exotic settings enriched by a color scheme of reds, mustards, icy blue and natural woods, guests sample such temptations as sushi, teppanyaki and Hong Kong-style barbecue, savoring Jia's distant ports of call just steps from the casino.

Top left: Dining room.

Top right: Banquette seating in dining room.

Right: Bar.

Opposite: Entrance with red waterfall.

Photography: Peter Malinowski/ InSite Architectural Photography.

Cleo Design

Beau Rivage Resort and Casino
High Limit Gaming Area
Biloxi, Mississippi

If the rich are different from everyone else, as F. Scott Fitzgerald observed, so are their accommodations. Consider the new, 5,136-square-foot High Limit Gaming Area, designed by Cleo Design, in the 1,740-room Beau Rivage Resort and Casino, in Biloxi, Mississippi, which has completely recovered from the devastation of Hurricane Katrina. Even before the storm, management had wanted to more clearly define the 151-seat High Limit Gaming Area so that its affluent guests would discern an appreciable difference in its ambience as part of a distinctive VIP experience. The new space fulfills the hopes of the Beau Rivage with an opulent yet informal interior that looks and feels different to customers at every step, from its high-profile façade, viewable from the hotel's entrance, to its slot room, cashier area, hospitality buffet, game tables, bar and host office. The façade is adorned by a geyser light incorporating glass crystal and accent lighting that is positioned at the entrance to draw the gaze of passersby directly inside. By contrast, the interior, resplendent in gold and silver mosaics, rich woods, upholstered walls, leather and mohair upholstery, and stately transitional furnishings updated with contemporary finishes, speaks of quiet comfort and unhurried luxury.

Left: Game tables.
Bottom left: Bar.
Bottom right: Lounge area.
Opposite: Slot room.
Photography: Peter Malinowski/ InSite Architectural Photography.

GAME KING
MULTI-GAME

Cleo Design

Las Vegas Hilton
Tempo
Las Vegas, Nevada

Top: Overall view of space.

Above: Detail of elliptical room divider.

Opposite: Close-up of seating area.

Photography: Peter Malinowski/ InSite Architectural Photography.

Constructed in 1969 and well established on its 80-acre site as a leading Las Vegas hotel, casino and convention hotel, the 3,174-room Las Vegas Hilton—the largest Hilton Hotel in the world as well as one of the largest hotel-casinos in town—perpetually renews itself with such lively additions as Tempo, a 5,205-square-foot "ultralounge," designed by Cleo Design. Situated in a previously little-used location where sheer walls and structural columns were uncovered in unexpected places, Tempo has been developed as a high-energy gathering place for a young, hip crowd. Its bar, table seating and lighting have been designed to give Tempo two contrasting identities, as a lively cocktail lounge by day and a sizzling nightclub with beverage service and performers by night. Contemporary appointments—featuring luminescent circular cocktail lounge tables and curvilinear seating, bold, vibrant colors along with stark black and white, and a battery of lighting sources—are all displayed against a novel backdrop of giant ellipses and circles that act as room dividers to define the otherwise open space. Thanks to its versatile design, Tempo shifts easily from one "tempo" to another, inviting its cocktail hour guests to return for an evening of great entertainment just hours later.

Cleo Design

MGM Grand
Studio 54
Las Vegas, Nevada

The pursuit of happiness has inspired the MGM Grand since the 5,044-room Las Vegas hotel and casino opened in 1993. In addition to its 170,000-square-foot Casino, it offers guests a wide choice of entertainment. Studio 54 represents one of the MGM Grand's most intimate attractions. This reinvention of Manhattan's legendary nightclub now features a splendidly remodeled entrance and VIP lounge, designed by Cleo Design. With DJs already drawing guests to fill the dance floor of the two-story, 4,926-square-foot space, the MGM Grand wanted a more exciting, upgraded entry experience and elite atmosphere in the VIP area.

The design solution heightens the sense of drama with strong graphics, cool contemporary and minimalist furnishings, a lush color scheme of raspberry, coral and dark neutrals, and spectacular lighting—introduced by custom fiber optic chandeliers at the entry made by weaving the fibers into intricate and graceful forms—suffusing such materials as taffeta and rich woods. Giving VIP guests a club within a club, the new facilities assures VIPs they've come to the right place.

Top: Entrance.

Right: VIP lounge.

Photography: Peter Malinowski/ InSite Architectural Photography.

Daroff Design Inc.

2121 Market Street
Philadelphia, PA 19103
215.636.9900
215.636.9627 (Fax)
www.daroffdesign.com

Daroff Design Inc.

Rittenhouse Hotel
Philadelphia, Pennsylvania

Left: Main lobby.

Below: Mary Cassatt Tea Room and Garden.

Bottom: Luxury suite living room.

Opposite: Grand ballroom.

Photography: Robert Miller.

Occupying the lower floors of a starkly modern, concrete-and-glass residential condominium tower overlooking Philadelphia's fabled Rittenhouse Square, the Rittenhouse Hotel flourishes as one of the most prestigious five-star hotels in the tradition-minded City of Brotherly Love. Providing impeccable service, elegant accommodations and Rittenhouse Square as its "entry garden," the hotel maintains some of Philadelphia's most spacious accommodations, featuring 98 guestrooms that include 13 lavish suites ranging in size from 450 to 600 square feet. The enthusiasm that business and leisure travelers feel for its amenities, which they actively share with condominium residents and others, is reflected in its AAA Five-Diamond rating—one of just three Pennsylvania hotels so honored. To maintain its high standards, the hotel recently retained Daroff Design to evaluate the condition of its interiors and recommend selective refurbishment and replacement of finishes and furnishings. Daroff Design's recommendations have reinvigorated the hotel's established eclectic/traditional decorative style with a fresh interpretation of traditional forms, design motifs and color palettes. Now that the extensive makeover of the entry lobby, ballroom, meeting rooms and pre-function areas has been completed, the travel industry is applauding with such honors as *Condé Nast*'s Traveler's Gold List of Best Places to Stay in the World.

Daroff Design Inc.

Parc Rittenhouse Condominiums and Club
Philadelphia, Pennsylvania

A grand 1920s-era former office tower facing Rittenhouse Square, the hub of Philadelphia's vibrant Center City residential community, has been triumphantly reborn as the new Parc Rittenhouse Condominiums and Club. With Rittenhouse Square just outside its front door and a spectacular Club penthouse with fitness center and swimming pool overlooking the Square, the Parc Rittenhouse showcases signature interiors designed by Daroff Design for its condominium residents. This comprehensive renovation comprises new interiors for the ground-floor entry lobby, elevator core, residence floor elevator lobbies and corridors, and space planning modifications to existing apartment layouts to produce more efficient and exciting living arrangements. Setting an example for the grand "club" style that characterizes the building's image, the sumptuous lobby has been designed with coved plaster ceilings, American

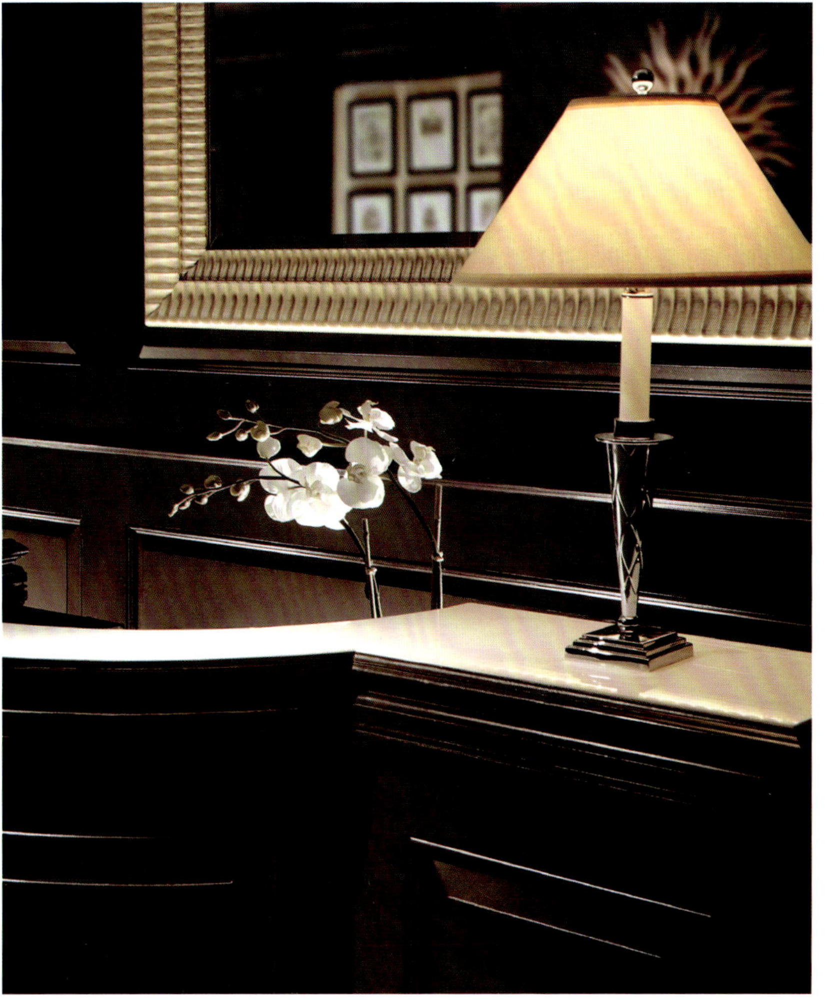

Walnut paneling, millwork and fireplace mantle, Santos mahogany, granite and marble inlay floor finishes, leather and wool upholstered custom furniture, accent carpets, traditional decorative and display lighting, artwork and accessories. In addition, Daroff Design has created luxury bathrooms and kitchens, selected a rich palette of apartment finishes, features and amenities, designed the sales and marketing center, and furnished three model apartments, assuring fortunate Philadelphians a warm welcome to their new homes in the Parc Rittenhouse.

Top: Armchair detail.
Left: Concierge desk.
Opposite: Lobby/lounge.
Photography: Robert Miller.

Daroff Design Inc.

Susanna Foo Gourmet Kitchen
Radnor, Pennsylvania

Celebrity chef Susanna Foo, James Beard Award winner and author, is credited with having invented Chinese/ French fusion cuisine, which helps explain why people along Philadelphia's Main Line are crowding into her new, 300-seat, 10,000-square-foot Susanna Foo Gourmet Kitchen restaurant in Radnor, Pennsylvania, designed by Daroff Design. The space reflects Foo's expanding media universe as well as her cuisine, revolving around an efficient and photogenic specialty kitchen where her gourmet cooking can be created daily and broadcast live on occasion throughout the restaurant and to PBS and Food Network audiences nationwide. Of course, the interiors function superbly for guests as well. Working as partners, Foo and Karen Daroff have created a sumptuous, front-of-house setting with a grand main dining room, more intimate dining areas, private party dining rooms, garden dining rooms and a take-out exhibition kitchen, enabling the facility to serve as both a local neighborhood restaurant and a regional destination for serious gourmets. The environment they've produced forms an indelible part of the guest experience, with dining areas unfolding in a sequential flow of spaces that is subtly articulated by such partitions as stone walls,

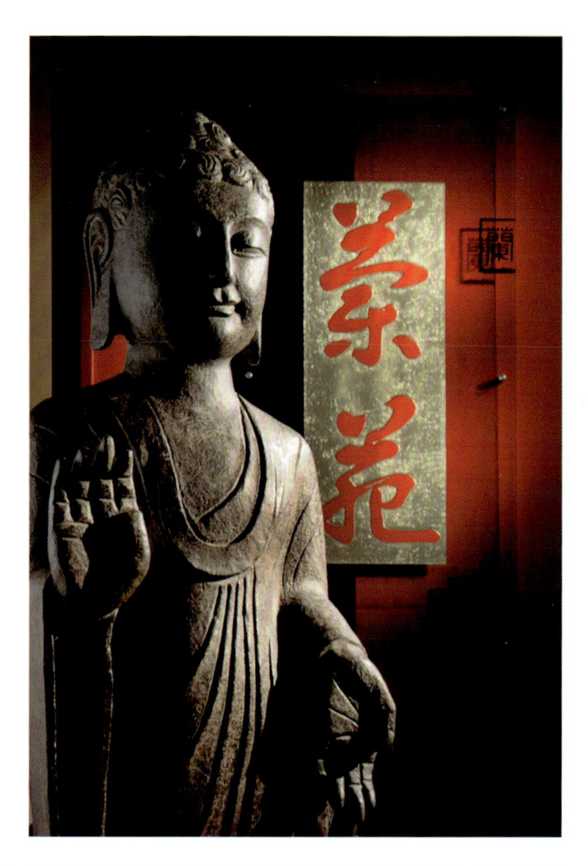

Left: Main dining area.

Bottom: Bar.

Opposite top: Buddha.

Opposite bottom: Hostess desk.

Photography: Robert Miller.

custom millwork screens, a "forest" of natural bamboo or a combination of intimate French-style banquettes and round dining tables. Foo's concern about the ingredients for her cuisine is also reflected in her interest in the materials and furnishings for her restaurant. Daroff has prepared a richly varied palette for her, blending an eclectic mix of authentic Chinese artifacts with distinctive contemporary furnishings and decorative elements from selected sources in China, highlighted by striking juxtapositions of natural stone, wood and bamboo materials against rich Chinese red walls, custom imported Chinese lanterns and shimmering hand-blown glass, and such exquisite details as a glistening stainless steel display kitchen, hardwood bar and wine display feature wall, traditional Chinese sculpture, kinetic video displays, and a Chinese gong sounded for guests' birthdays. Susanna Foo Gourmet Kitchen is thus the perfect place to sample such signature entrées as panko-crusted crab cakes or honey walnut chicken with mango and ginger, where guests can savor the cuisine and décor together as one delicious experience.

Left: Main Dining Area.

DAS Architects

1628 JFK Boulevard
Philadelphia, PA 19103-2125
215.751.9008
215.751.9118 (Fax)
www.dasarchitects.com
info@dasarchitects.com

DAS Architects

Rae

Philadelphia, Pennsylvania

Hungry power brokers arriving at Amtrak's 30th Street Station in Philadelphia can now join local business people for serious dining next door at the Cira Centre, thanks to the new 220-seat, 12,500-square-foot Rae restaurant, designed by DAS Architects for noted chef Daniel Stern. In setting the stage for such "New American" fare as rack of lamb with ricotta and feta soufflé or sauteed halibut with tomatoes and herbs, the design deftly matches the energy and style of its location, the building's soaring modern lobby, featuring pedestrian bridges and skyline views of Center City. Rae engages several interconnected areas. While an island-style finishing kitchen anchors a main dining area, bar, lounge and glass-enclosed bakery, a mezzanine staircase at the end of a booth-lined main circulation spine rises to a glass-enclosed wine room and private dining rooms. Nearby, a 200-seat ballroom, preparatory kitchen, and back-of-house space complete the facility. The crisp, modern interior, comprising limestone, plaster, stainless steel, cast gypsum, art glass, anigre and maple woodwork, classic furniture in mahogany, leather and textiles, and sophisticated lighting, has created a space within a space for Rae to delight customers and win *Philadelphia* magazine's "Best New Restaurant for 2007" citation. Also, Rae was named one of the best new restaurants in America, 2007 by *Esquire*.

Left: Skyline view from bar counter.

Below left: Wine room.

Below right: Lobby dining.

Opposite top left: Mezzanine view.

Opposite top middle: Banquette seating and open finishing kitchen.

Opposite top right: Lounge with bar beyond.

Far left: Bar.

Photography: Peter Paige, Jeff Totaro.

DAS Architects

Salt Creek Grille
El Segundo, California

Of course Frank Lloyd Wright never dined at Salt Creek Grille, a chain of classic American grill restaurants founded in 1996 and housed in freestanding Craftsman-style wood buildings amidst naturalistic landscaping. Yet the handsome new, 300-seat, 10,000-square-foot Salt Creek Grille, designed by DAS Architects, at Plaza El Segundo, a trendy shopping center in El Segundo, California, acknowledges Wright in numerous, subtle ways. Its single-story structure houses a lobby, bar, lounge, main dining room, small dining room, wine display and outdoor bar and lounge within a dynamic, open environment of sweeping horizontal lines. What clearly gratifies diners is the way such popular dishes as grilled pork chop (brined in apple cider and double-thick cut) and Mesquite grilled Atlantic salmon are complemented by a setting where no detail large or small is overlooked. From the cathedral ceiling overhead to such individual objects as the finely crafted bar, timeless wood furniture, custom pendant lights and wall sconces, and gray blankets bearing the restaurant's logo that invite diners to linger among the outdoor gardens and fire pits, the contemporary architecture consistently projects a warmth and pride of place—along with the Salt Creek Grille brand—that has diners coming back for more.

Above: Main entrance.

Right: Exterior with outdoor gardens.

Below: Main dining room, bar and lounge.

Opposite top left: Small dining room.

Opposite top right: Bar.

Photography: Peter Paige.

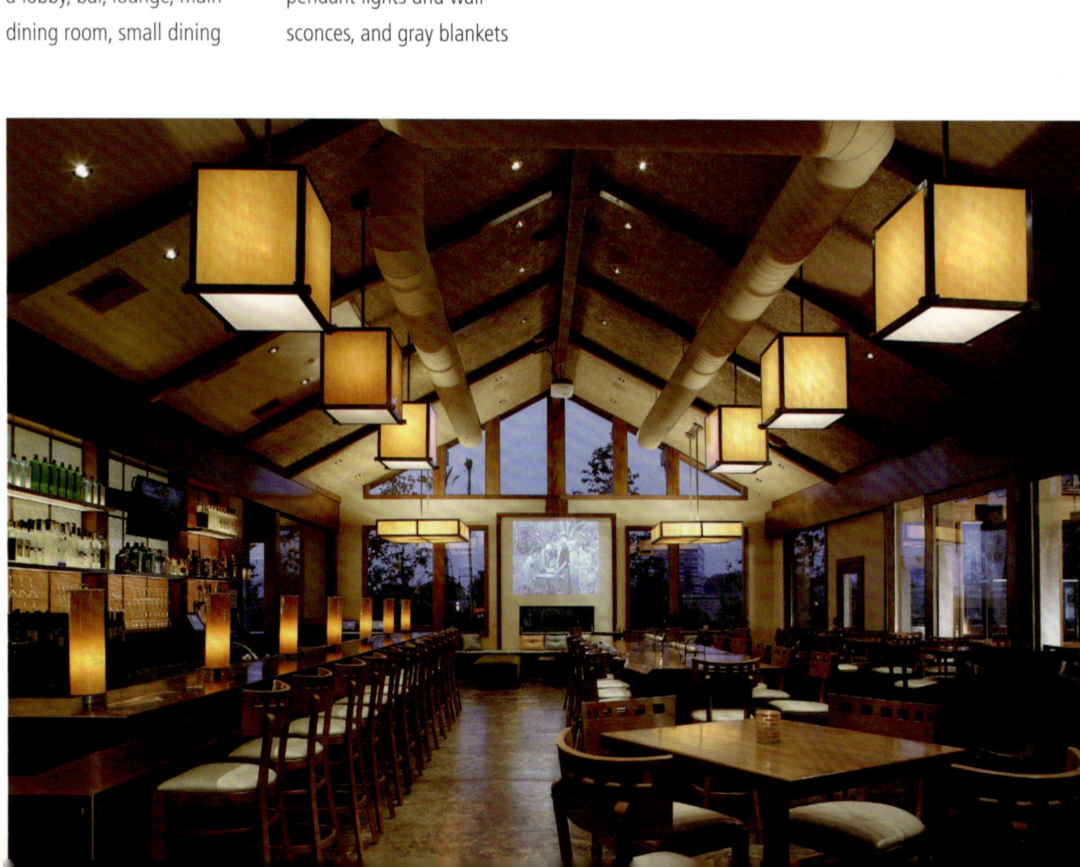

DAS Architects

Le Bec-Fin
Philadelphia, Pennsylvania

When Georges Perrier, chef and owner of Le Bec-Fin, a legendary French restaurant serving Philadelphia for 35 years, returned from Paris to behold his remodeled facility, he wept over its beauty.
It's easy to see why. The comprehensive renovation, designed by DAS Architects, has transformed the lobby, 74-seat main dining room, 16-seat small dining room, wine display and 24-seat private dining room, delighting loyal clients and new, younger guests alike. By opening up the long, narrow (25-foot-wide) space, introducing lighter finishes, and painstakingly recreating the architecture of a 19th-century Parisian salon, the design lets Philadelphians rediscover Le Bec-Fin, savoring the intimacy of the small dining room at the front, and marveling over the splendor of the main dining room and its high ceiling, chandeliers, and open stairway to the private dining room at the back. Materials and furnishings were as lovingly chosen as the restaurant's sauteed black sea bass with watercress and red radish fricassee or filet of beef with Bordelaise sauce, endive and walnut fricassee. Retaining the French chandeliers from the original interior, the design has added custom millwork, cast plaster with gold leaf, silk fabrics, custom carpet, antique mirrors, and custom tables and chairs. Bon appétit, Philadelphia!

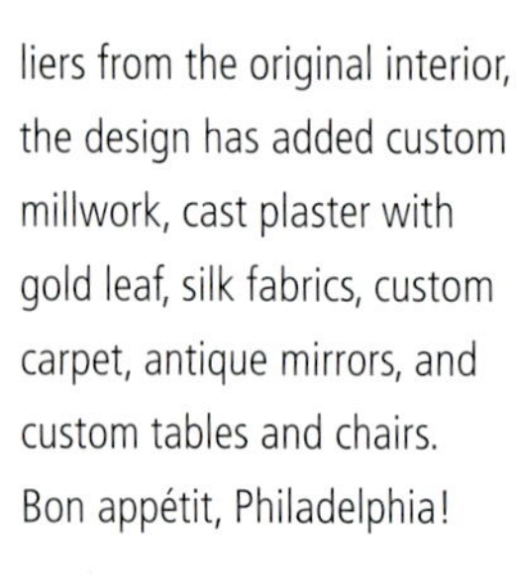

Top left: Main dining room and stairway to private dining room.

Top right: Fireplace in main dining room.

Above right: Small dining room.

Right: Silk fabric covered wall panels.

Opposite: Doorway between main and small dining rooms.

Photography: Peter Paige.

DAS Architects

Wolfgang's Steakhouse
New York, New York

Like a good writer honoring the credo, "Write what you know," Wolfgang Zweiner, retired head waiter of Peter Luger's, New York's perennially top-rated steakhouse, followed the wisdom of four decades with his former employer by opening his own midtown Manhattan steakhouse in 2004 where the Vanderbilt Hotel once reigned. Now, the original Wolfgang's Steakhouse has done so well that Zweiner has opened a second one, designed by DAS Architects, in the chic Tribeca neighborhood. Once again, the new, 195-seat, 7,500-square-foot Wolfgang's draws on proven strategy by transplanting the richness of the original restaurant's historic tiled, Gustavino-vaulted interior to a raw industrial shell. Accommodating a bar, lobby, main dining room, wine room, private dining room and kitchen, the design replicates the vaults through floating arched ceiling planes finished in handmade glass mosaic tile, and adds appropriate colors and furnishings to effectively capture the original restaurant's turn-of-the-century masculine mood. However, the new Wolfgang's wisely uses the three connecting industrial bays in its space to place the main entrance, bar, wine display and casual table seating in the middle bay; main dining room in one end bay, and private dining room and bathrooms in the other end bay. Historic or not, it works.

Top right: Private dining room.
Right: Main dining room.
Bottom: Bar and wine room.
Photography: Peter Paige.

DiLeonardo International, Inc.

2348 Post Road
Suite 501
Warwick, RI 02919
401.732.2900
401.732.5315 (Fax)
www.dileonardo.com

DiLeonardo International, Inc.

Sheraton Hotel Bangalore
Bangalore, India

What's ahead for India's booming economy? The possibilities can be glimpsed in the "Garden City of India," Bangalore, capital of the state of Karnataka, fashion center, home to some six million residents, and now "India's Silicon Valley." One hotel likely to attract today's business travelers streaming into India's seventh largest city is the new 345-room Sheraton Hotel Bangalore, currently under construction with an interior design by DiLeonardo International. Its sleek, contemporary facilities, including a lobby, lounges, restaurants, bars, ballroom, and executive lounge as well as guestrooms, reflect the needs of the fashion and high-technology worlds. Using the building's distinctive curving shape to direct circulation, the design will feature a spacious lobby with specific zones marked by flooring, lighting and walls to host fashion shows, a monumental stair that serves as a focal point for special functions, and a variety of public spaces that take their aesthetic cues from leading fashion designers to combine contemporary furnishings, sophisticated neutral colors with key color accents, and such timeless materials as bronze, silver, limestone and marble. Yet looks won't be everything. Though high-tech hardware and services will be largely invisible, business people will instantly detect their presence in this hotel for India's "Silicon City."

Top left: Guestroom.
Above left: Restaurant.
Above: Lobby lounge.
Opposite left: Lobby.
Opposite right: Bar.
Illustration: Courtesy of DiLeonardo International, Inc.

DiLeonardo International, Inc.

InterContinental Lagos
Victoria Island, Lagos, Nigeria

Oil-rich Nigeria is an African nation of over 115 million people with multiple languages, cultures and religions. Lagos plays an important role as its largest city and commercial center, enabling the 358-key InterContinental Lagos, now under construction on Victoria Island, to make a significant contribution to business and social life. The location of the hotel—where DiLeonardo International is designing the porte cochere, lobby, lobby bar, three restaurants, meeting rooms, auditorium, boardroom, business center, banquet hall, retail stores, spa, gym, swimming pool, Zen garden, and guestrooms—reflects the geography of Lagos, whose 12 million residents are distributed across an archipelago in a lagoon and the adjoining mainland. One of three large islands in the archipelago, Victoria Island shelters much of the nation's ruling elite, a vibrant commercial district with shopping malls and movie theaters, as well as a large foreign expatriate community. To welcome guests arriving for business and social events—including many anticipated weddings—the interiors are being designed as bright, airy and dynamic spaces outfitted with oversized and comfortable contemporary furnishings in festive colors using stone, wood and other natural materials. If the liveliness of Nigerian weddings offers any indication, the InterContinental Lagos is going to be a very congenial place.

Right: Lobby lounge.

Below: Lobby reception.

Below right: Presidential suite.

Below middle right: Club suite.

Bottom right: All-day restaurant.

Illustration: Courtesy of DiLeonardo International, Inc.

DiLeonardo International, Inc.

Radisson Medeu Resort, Kazakhstan
Almaty, Kazakhstan

Off the beaten track as the ancient settlement of Almaty may be, tracing its roots to the third century B.C., proven oil reserves have landed Kazakhstan's former capital and largest city squarely on the global energy map.

The city of some two million citizens is not totally focused on the petroleum business, to be sure. The new Radisson Medeu Resort, a 155-room hotel now under construction, will feature a luminous interior design by DiLeonardo International comprising a lobby, lounge, all-day restaurant, specialty restaurant, ballroom, guestrooms and presidential suite, with the goal of pampering guests visiting the picturesque Medeu Valley, some nine miles from Almaty. Because people treasure the Medeu Valley for its beautiful, natural landscape and excellent skiing, the Radisson Medeu Resort will surround them with clean-lined, straightforward modern design that subordinates itself to the stunning scenery just beyond the windows. Yet there will be reminders of nature's splendors in such design details as the water features, slates and marbles, glass ceilings, richly colored, embossed leathers and plush

carpets. In this manner, the Radisson Medeu Resort should be able to remind guests that happy memories of their stay in the Medeu Valley will begin and end with its hospitality.

Top right: Specialty restaurant.
Right: Ballroom.
Below: Lobby.
Opposite left: Lobby lounge.
Opposite bottom left: All-day restaurant.

DiLeonardo International, Inc.

Hyatt Regency
Newport, Rhode Island

Above: Restaurant.

Left: Lobby bar.

Below left: Spa.

Bottom left: Guestroom.

Illustration: Courtesy of DiLeonardo International, Inc.

Improving on the past is not something undertaken lightly in Newport, Rhode Island, a resort community founded in 1639 where magnificent, turn-of-the-century "cottages" designed by the likes of McKim Mead & White still hold court. But the respected Hyatt Regency Newport, a deluxe, AAA 4-Diamond hotel located on Goat Island, grew haphazardly over many decades, so an elegant and timeless updating of both the architecture and interior design by DiLeonardo International is coming as a welcome change for the 365-bed hotel. When the remodeling now underway is complete, covering the registration area and lobby, restaurant, ballroom, meeting rooms, spa, guestrooms and suites, guests will discover a heretofore missing continuity of design vision and detailing that befits a luxury resort experience. Of course, the imagery of the new design draws inspiration from Newport's legendary nautical environment to establish a modern, casual setting of contemporary furnishings and local materials like granite and slate. After all, a resort community approaching its fourth century must be doing something right.

Haverson Architecture and Design

63 Church Street
Greenwich, CT 06830
203.629.8300
203.629.8399 (Fax)
www.haversonarchitecture.com

Haverson Architecture and Design

The Griswold Inn Wine Bar
Essex, Connecticut

Travelers fond of charming historic places and hungry for the latest culinary delights can now find them in one timeless, if newly remodeled, establishment: The Griswold Inn Wine Bar, in Essex, Connecticut, designed by Haverson Architecture and Design. Established in 1776, the Griswold Inn is a beloved New England icon proffering period architecture, traditional regional fare, sea shanty sing-alongs and jazz concerts, and museum-quality maritime art. The 3,500-square-foot Wine Bar, offering over 50 wines by the glass and an innovative small plate menu, skillfully updates the "Gris" without diminishing its authenticity. Taking a former dining room and its 18-foot seascape mural, the design introduces a more expansive, contemporary ambiance in the two-part, 2,000-square-foot public area through an uncommon seating arrangement, whereby a bar, banquettes and table seating encircle a raised common table shared by friends or congenial strangers. The scheme is fitted out with finely crafted mahogany trim and oak flooring, warm bronze nautical lanterns used as pendant lights, and ceiling coves in each dining space, equipped with concealed lighting and belt-driven, turn-of-the-century fans to "lift" the low ceiling. As enthusiastic restaurant reviewers have reported, the Wine Bar makes the "Gris" feel 230 years young.

Top: Hand-carved sign.

Above: Essex room.

Left: Common table.

Opposite: Wine bar.

Photography: Peter Paige Photography.

Haverson Architecture and Design

Junior's Times Square
New York, New York

Left: Entrance and bar as seen from dining room.

Below: Exterior.

Bottom right: Takeout.

Opposite: Bar.

Photography: Peter Paige Photography.

Delivering prize-winning cheesecakes from a venerable Brooklyn restaurant to Times Square, Manhattan's legendary theater district, is no big deal. However, transplanting the spirit of the restaurant to Times Square's Shubert Alley, a mid-block pedestrian walkway connecting 44th and 45th Streets between Broadway and Eighth Avenue, is another matter. That a branch of Junior's is thriving among Broadway's theaters affirms the validity of its 200-seat, 7,500-square-foot design, created by Haverson Architecture and Design. Previous work for the third generation owners, brothers Alan and Kevin Rosen, and their father, Walter, to design Junior's Café Takeout and Dessert Shop at Grand Central Terminal and renovate the original Brooklyn restaurant, let Haverson analyze the restaurant's customer experience. Accordingly, the new Junior's sandwiches an L-shaped dining room of tables, banquettes and a bar between a standalone bakery and a takeout area, providing views of 45th Street and Shubert Alley, incorporating dramatic lighting, and seasoning the interior spaces and outdoor café with Junior's memorabilia, vintage photographs of Brooklyn landmarks, and such Junior's motifs as its signature orange and white striped canopy, familiar logo and signage, and early 1960s interior design. What better place for pre-theater dinner or post-theater supper—with Junior's fabulous cheesecake for dessert?

Haverson Architecture and Design

Westin Stamford
Lobby and Senses Restaurant
Stamford, Connecticut

Savvy hoteliers are discovering how a good lobby restaurant, bar and lounge, banquet facilities and meeting rooms can generate lucrative business from local residents as well as out-of-town guests. When HEI Hospitality acquired the Westin Stamford, a popular destination for business travelers, conventions and private parties nestled in a park-like setting beside I-95 in Stamford, Connecticut, it retained Haverson Architecture and Design to renovate these facilities—7,000 square feet of space overall—as part of an updating of the hotel. The project lets guests relax and unwind from the stresses of travel and business, employing natural materials and a garden motif to lure them into a soothing oasis for drinks and exquisite Eurasian cuisine. What gives the design its undeniable appeal is an organic approach that touches every last detail. The lobby lounge, for example, is finished in slate flooring

and crafted wood furniture. Similarly, the Senses features such highlights as hourglass-shaped, wood sheathed columns topped with dynamic lighting displays, space dividing planters bearing bamboo shoots and river stones, numerous direct and indirect lighting sources, and elegant contemporary furnishings. In a favorable appraisal of the facelift, the Stamford Advocate's reviewer confessed, "I felt like I arrived at the wrong hotel!"

Left: Bar.

Top right: Senses dining room.

Middle: Lounge.

Bottom: The Cave, a private dining room.

Photography: Peter Paige Photography.

Haverson Architecture and Design

Mohegan Sun Hotel and Casino
Uncas American Indian Grill and Chief's Bagels
Uncasville, Connecticut

It's obvious that food service is part of the fun at Mohegan Sun Hotel and Casino, a Native American casino in Uncasville, Connecticut. The physical environments designed for two recently opened eateries by Haverson Architecture and Design, Uncas American Indian Grill and Chief's Bagels, reveal how design not only facilitates efficient operations, it helps build unique experiences and customer loyalty. In Uncas American Indian Grill, regional American cuisine indigenous to the Mohegan Territory of Eastern Connecticut is served in a warm, rustic, open-hearth environment comprising a 340-seat dining room, 25-seat bar and 40-seat private dining room. Modeled like a clearing in a forest, the Grill combines theatrical lighting, such natural materials as fieldstone, cherry wood, and decorative tile, and natural birch trees crossing before an illuminated wall to invite guests to observe entrees prepared over a wood fire and rotating spit grill. Likewise, Chief's Bagels serves homemade bagels, pastries and freshly brewed coffee against the backdrop of a spiraling, wood-burning oven, which guests can watch from organically shaped, internally illuminated banquette seats. Both eateries are meant to be noticed. Agrees Gary Crowder, Mohegan's senior vice president of food and beverage, "These are among our most spectacular dining outlets."

Top left: Chief's Bagels.

Top right: Uncas dining room.

Bottom right: Uncas entry and bar.

Photography: Peter Paige Photography.

HBA/Hirsch Bedner Associates

3216 Nebraska Avenue
Santa Monica, CA 90404
310.829.9087
310.453.1182 (Fax)
www.hbadesign.com

HBA/Hirsch Bedner Associates

The Ritz-Carlton Beijing, Financial Street
Beijing, China

Right: Lobby stair and reflecting pool.

Below: Corridor.

Opposite: Greenfish Restaurant dining room.

Photography: Sun Xiangyu (Greenfish Restaurant, corridor), Leo Ying Li (lobby).

Dealmakers and vacationers aren't waiting for the 2008 summer Olympic Games to visit Beijing, China's legendary capital and second largest city. With so much already happening in the capital of a country that will soon overtake Germany as the world's third-largest economy, the opening of the new, five-star, 253-room, 266,900-square-foot Ritz-Carlton Beijing, Financial Street, featuring interior design by HBA/Hirsch Bedner Associates, could not be timed better. The luxury hotel's sleek, glass-and-chrome tower stands at the intersection of Taipingqiao and Jinchengfangdong, the heart of Beijing's Financial District, offering guests elegant, state-of-the-art facilities that include three restaurants, tea parlor and bar, spa, health club and fitness center, ballroom, three meeting rooms, prefunction and function space, and such additional amenities as a 24-hour business center, salon and gift shop, along with luxurious guestrooms and suites. Its suave, sophisticated interiors showcase a classic, contemporary style accented with an unmistakable touch of Chinese traditional design. Universal acclaim for these world-class accommodations reflects

considerably more than the design, of course. For HBA/ Hirsch Bedner Associates, coordinating the planning, design, fabrication and installation with local suppliers and design institutes represented a particularly satisfying accomplishment. Bridging gaps in knowledge and experience between designer and local partners required an educational effort and unflagging diligence at every step. The effort has paid off spectacularly with higher quality and lower cost in spaces as large as the ballroom, as intricate as Cèpe, the Italian restaurant, and as intimate as the typical guestroom. At Cèpe, for example, black lacquered open shelves displaying exquisite Chinese porcelain act as room dividers to create a memorable setting for northern Italian cuisine. The 5,425-square-foot ballroom, the focal point of 10,667 square feet of meeting space, displays 35 Murano-blown glass peony flower-shaped chandeliers and Liu Li colored glass inlays in ceiling panels as a glamorous backdrop for events. Hand-crafted, residential-style contemporary furnishings with Oriental accents make guestrooms feel warm and inviting even though they are the most spacious—starting at 538 square feet—in Beijing's luxury hotel tier. If this environment sounds equally appealing to leisure travelers, it should. Such popular attractions as the Forbidden City, Temple of Heaven, Beijing Zoo and Summer Palace are within a convenient walk or drive from the doors of the Ritz-Carlton Beijing, Financial Street.

Right: Qi Restaurant private dining room.

Far right: Cépe Restaurant.

Opposite top left: Suite master bath.

Opposite top right: Spa.

Photography: Leo Ying Li (Cépe Restaurant, Spa), Sun Xiangyu (Qi Resturant, Master bath).

Wine
BOTTICELLI

HBA/Hirsch Bedner Associates

Pan Pacific Seattle
Seattle, Washington

Call it a neighborly way to enjoy Seattle. Situated in the South Lake Union neighborhood, the new, 160-room Pan Pacific Seattle, with interiors designed by HBA/Hirsch Bedner Associates, shares neighbors' goods and services with guests. Because the Pan Pacific Seattle is part of 2200 Westlake, a 550,000-square-foot mixed-use project, it is just steps from 261 condominium homes, two signature restaurants, 39,000 square feet of retail shops, Starbucks, Whole Foods Market and a 16,000-square-foot landscaped plaza, along with its own grand lobby, guestrooms, 5,500 square feet of meeting space, including a 1,674-square-foot ballroom, lobby bar and spa. Naturally, the hotel assures guests a satisfying experience within its own stylishly contemporary "East meets West" environment, comprising Zebrawood millwork and furnishings, Jerusalem limestone floors and accents, fine textiles and carpet, and distinctive lighting fixtures. The Grand Lobby, for example, boasts a dynamic setting characterized by a high ceiling, cantilevered elliptical staircase and mobile-like chandelier. By contrast, guestrooms foster privacy and comfort through such luxurious amenities as Hypnos beds, Egyptian cotton bedding, Herman Miller chairs and other classic furnishings, soaking tubs and the ultimate benefit of a good neighborhood: inspiring views, which place the Space Needle and downtown Seattle front row and center.

Top left: Denny suite.

Top right: Junior suite bath.

Above: Lobby bar.

Left: Grand Lobby.

Opposite top left: Lakefront Ballroom.

Photography: Stephen Allard.

HBA/Hirsch Bedner Associates

Okada Restaurant
Wynn Resort Las Vegas
Las Vegas, Nevada

Sushi may never become as familiar to Americans as hamburgers, but consider Okada, a new, 271-seat Japanese restaurant, designed by HBA/Hirsch Bedner Associates, at the Wynn Resort Las Vegas, in Las Vegas. Here, in a timeless interior where Japanese modern and traditional design converge, diners are gathering for authentic teppanyaki, traditional sushi, and modern Japanese creations. Though chef Masa Ishizawa and his staff deserve full credit for their culinary accomplishments, they receive strong support from the space, which is visually pleasing and easy to navigate. The design contrasts traditional materials such as maple wood paneling, bamboo lattices, stone walls, silk wallcoverings and iron gates with such obvious symbols of the industrial world as acrylic panels and steel bolts to establish a contemporary look interwoven with historic references. Since the restaurant's teppanyaki cooking requires a distinct physical setting, the design skillfully differentiates the overall space into smaller areas without partitioning it. Design details are as delicate as the cuisine: a wooden canopy over the robatayaki counter, leather straps holding massive wood beams, origami-like ceiling planes of acrylic paneling, and more. From the entrance through the bar to the dining, private dining and teppanyaki areas, Okada is a feast for the eye.

Top: Dining room with waterfall view.

Above: Robatayaki counter.

Left: Teppanyaki room.

Opposite: View towards bar.

Photography: Erhard Pfeiffer.

HBA/Hirsch Bedner Associates

The Regent Beijing
Beijing, China

Although historic China is now yielding rapidly to modernization, guests of the new, 500-room Regent Beijing, featuring interior design, project management and art consultation by HBA/Hirsch Bedner Associates, greets old and new Beijing every day. Strategically positioned between the modern business district and venerable *hutongs* just steps from the Forbidden City and Tiananmen Square, the award-winning Regent Beijing offers a rich, memorable experience fusing northern Chinese tradition with modern luxury living and museum-quality furniture and other furnishings. Clear, basic circulation encourages guests to enjoy the hotel's grand public areas and such amenities as five restaurants and bars, spa, health club and extensive meeting facilities, highlighted by a 6,730-square-foot Ballroom. These ambitious interiors are beautifully detailed through superior craftsmanship and such timeless materials as limestone, mahogany and leather. At Lijing Xuan Chinese restaurant, for example, rare Chinese red sandalwood cabinetry and millwork produce a traditional environment that is warm and opulent. The 443 guestrooms and 57 suites, by contrast, season their sophisticated, contemporary style with such accents as Chinese occasional furniture and a marble bathroom with a glass wall that looks out to the bedroom. In the seat of China's emperors, The Regent Beijing's guests are very much at home.

Top left: Lijing Xuan Restaurant.

Top right: Exterior.

Above middle right: Ballroom.

Right: Guestroom.

Opposite top: Crescent Lounge.

Opposite bottom left: Daccapo Italian restaurant.

Opposite bottom right: The Bar and Grill.

Photography: George Mitchell and others, courtesy of The Regent Beijing.

HBA/Hirsch Bedner Associates

Wynn Macau
Macao, China

Top left: Logo Apparel.
Top right: Casino.
Above left: Red 8 Restaurant.
Left: Okada Japanese restaurant.
Below left: VIP lounge.
Opposite: VIP Registration.
Photography: Erhard Pfeiffer.

Steve Wynn's innovative approach to the gaming experience—witness the spectacular Wynn Las Vegas—has now struck gold on the Cotai Strip in Macao, China. At the new, 600-room Wynn Macau, featuring public areas designed by HBA/Hirsch Bedner Associates, guests revel in an exuberant destination where gaming is part of a larger experience. The action never stops because the glamorous 100,000-square-foot casino is complemented by a monumental registration area, six stylish gourmet restaurants and numerous smaller restaurants and cafés, exciting entertainment lounges, state-of-the-art conference facilities, stately spa and health club, and 26,000-square-foot luxury retail promenade. The design excels in giving each activity an original and memorably beautiful setting. For the casino, for example, a high, coffered ceiling, exquisite drapery, elegant Italian crystal chandeliers, richly patterned carpet and intricate mosaic tile flooring establish an unmistakable allure. But there's nothing ordinary either about the Wing Lei Chinese restaurant, where guests enter a moon gate to marvel over a flying dragon of 80,000 crystals hovering above the main dining room, or Okada Japanese restaurant, displaying giant sake barrels crafted by a professional boat builder in a striking, contemporary space. The Wynn Macau's overwhelming popularity is reflected in its just completed second phase expansion.

HBA/Hirsch Bedner Associates

Caesars Palace
Concierge Suite
Las Vegas, Nevada

Reenacting the glory that was Greece and the grandeur that was Rome in the irrepressible American way, Caesars Palace has reigned as one of the world's best-known resort casinos from its 85-acre site on the Las Vegas Strip since 1966. Despite a bounty of 3,348 rooms, 26 restaurants and cafes, 4.5-acre Garden of the Gods, extensive spa and salon, 240,000 square feet of meeting space, and 4,100-seat Coloseum, it keeps inventing new indulgences for guests. One of the latest is the two-bedroom, 2,070-square-foot Concierge Suite, designed by HBA/ Hirsch Bedner Associates. Comprising an entry foyer with powder room and closet, double-height living/ dining area, master bedroom with five-fixture bath and dressing area, and double queen guestroom with four-fixture bath and dressing area, the Concierge Suite caters to the demanding individual seeking the ultimate in spacious, multi-use entertainment accommodations for three to four guests. The airy, elegant space is the result of demolishing an existing concierge/reception area in Caesar's Tower Suites and reshaping the volume with modern architectural interiors of mahogany, walnut, leather, linen, mohair and marble, luxury bath fixtures, classic contemporary furnishings, and custom residential-style lighting. Maybe Imperial Rome can't be recreated in a day, but the Concierge Suite surely helps.

Upper left: Living/dining area.
Lower left: Master bedroom.
Photography: Erhard Pfeiffer.

HBA/Hirsch Bedner Associates

Pudong Shangri-La Chi Spa
Shanghai, China

Above: Corridor.

Left: Couples suite.

Photography: George Mitchell.

Was anyone surprised that travelers in Shanghai quickly discovered the beauty, convenience and comfort of the Pudong Shangri-La, providing five-star, luxury accommodations and service in the heart of the dynamic Pudong business district? Demand has been so strong the hotel recently opened a new, 36-story, 375-room Grand Tower, raising total room count to 981 and giving more guests access to its restaurants, lounges, bars, health clubs, indoor lap pools, ballrooms, business center and other facilities. In the midst of so much activity, the Himalayan-themed, 8,611-square-foot Chi, The Spa at Shangri-La, designed by HBA/Hirsch Bedner Associates, has quickly become one of the most popular new offerings. Sophisticated yet casual, the award-winning spa consists of an entry/reception area and stair, elevator lobby and circulation spaces, treatment rooms, specialty suites (wet and beauty), couples suites and male/female facilities for hydrotherapy, sauna and steam zone. The design sets a quiet, spiritual mood for guests through an array of intimate spaces defined by such Tibetan-style elements as water walls, screens and niches, plus basic materials like stone, timber, metal and textiles, and contemporary furnishings incorporating Asian motifs. That Frommer's praises Chi's environment as "exquisite" may be all business and leisure travelers need to know.

HBA/Hirsch Bedner Associates

Jumeirah Carlton Tower
Gilt Champagne Bar
London, United Kingdom

You needn't be a visiting Asian or Arab head of state in London to surrender to the allure of the Jumeirah Carlton Tower, located at Cadogan Place and Sloane Street, close to the city's most fashionable shopping areas. Of course, VIPs are often drawn to the hotel's 220 guestrooms and 59 suites, attesting to its lavish and recently remodeled accommodations as well as its convenient site. Yet anyone can enjoy such attractions as the delightful new, 30-seat Gilt Champagne Bar, designed by HBA/Hirsch Bedner Associates. Inspiration for the design came from the late 1940s, when more fluid and curvaceous forms succeeded Art Deco and Art Moderne, and glamour was in vogue again following World War II. Positioned at the end of the hotel's central corridor, the space focuses on a gemlike bar with a beveled, mirrored edge, an oversized crystal champagne cooler illuminated from below, and a chandelier designed to evoke the effervescence of champagne. Whether guests are seated at the bar or at tufted banquette seating framed by upholstered leather walls and surrounded by period-style furnishings immersed in a champagne-inspired palette, they become part of a look that brings the spirit of Chanel, Balenciaga and Norell vividly to life.

Top left: Banquette.

Top right: Entrance.

Right: Bar and custom chandelier.

Photography: Courtesy of Jumeirah Carlton Tower.

HKS Hill Glazier Studio

1919 McKinney Avenue
Dallas, TX 75201
214.969.5599
214.969.3397 (Fax)
www.hksinc.com

HKS Hill Glazier Studio

Four Seasons Resort Hualalai
Kona, Hawaii

You won't find the standard big central building or guestrooms sharing long corridors at the Four Seasons Resort Hualalai in Kona, on the Kona-Kohala Coast of the Big Island of Hawaii. This 243-room resort hotel, featuring architecture designed by HKS Hill Glazier Studio, salutes Hawaii's golden age with a village of two-story, bungalow-style guestrooms arrayed like a crescent to form a *kipuka*, a landscaped oasis amidst lava flows. Among its original features are a lobby, restaurants, and spa designed as open-air pavilions with unobstructed views of the Pacific Ocean, a golf clubhouse overlooking the 18th fairway of a Jack Nicklaus-designed golf course, five pools, and guestrooms clustered around four informal courtyards and endowed with ocean views, private terraces, open-air bathrooms, and bedrooms and living rooms appointed in traditional Hawaiian-style furnishings. According to John Hill, AIA, a principal of HKS Hill Glazier Studio, "The residential scale mahogany and cedar construction of the Four Seasons reflects the design team's conviction that the result would appeal to the couples and small groups expected as guests and delight travelers seeking alternatives to conventional accommodations." Echoing his sentiments, Frommer's declares, "This is a great place to relax in the lap of luxury."

Top: Ocean view from guestroom.

Above: Restaurant lanai.

Opposite top left: Guest suite.

Opposite top right: Guest bath.

Right: Entry lobby.

Photography: Robert Miller Photography.

HKS Hill Glazier Studio

W Dallas-Victory Hotel & Residences
Dallas, Texas

This Cinderella story, set in Victory Park, a 72-acre brownfield site adjacent to American Airlines Center in Dallas now emerging as a mixed-use commercial development with a 12 million-square-foot build-out, stars a new icon on the Dallas skyline: the W Dallas-Victory Hotel & Residences. This 33-story addition to Starwood's hip boutique hotel chain was designed by HKS Hill Glazier Studio to combine a 252-room hotel with a 150-unit luxury condominium, letting residents share hotel amenities with guests. Though the design vigorously extends the W brand, it also endorses the chain's philosophy of having W Hotels provide unique experiences. "We've brought a fresh look to downtown," notes Eddie Abeyta, AIA, a principal of HKS Hill Glazier Studio, which collaborated with interior design firms handling such spaces as the lobby, restaurant, retail space, and conference center on the first two floors, the pool, spa, and fitness facility on the 16th floor, and the bar and outdoor terrace on the 33rd floor. From the bold, modern architecture—using the pool to mark the transition from hotel to condominium and mimicking Tennessee limestone on the façade with warm gray concrete—to the warm, open, and light-flooded interiors, the design fulfills Big D's motto to "live large, think big."

Above left: Lobby.
Above right: Hotel pool.
Bottom: Skyline with hotel.
Opposite: Façade.
Photography: Blake Marvin.

HKS Hill Glazier Studio

The One&Only Ocean Club
Paradise Island, The Bahamas

Old should look old and new should too, when guests register at The One&Only Ocean Club, on Paradise Island, The Bahamas. Such is the power of tradition among the legendary hotel's affluent clientele—as well as the power of architecture from HKS Hill Glazier Studio as the designer of the recently completed expansion—that no visual inconsistencies mar the historic patina. Since the Ocean Club opened in 1963 on the former estate of Huntington Hartford, it has epitomized the British colonial style of architecture known as Old Nassau.

However, HKS Hill Glazier Studio was enthusiastically encouraged to update the facilities through the discreet addition of 50 new guestrooms and three 5,000-square-foot villas, a new, 25,000-square-foot entry building, a restaurant, bar and dining terrace, and a new colonnade around the pool. "The client wanted our improvements to be invisible," recalls Bob Glazier, AIA, a principal of HKS Hill Glazier Studio. "Fortunately, this is easier than preserving historic architecture." In fact, the expansion greatly enhances the guest experience. Guests enjoy such amenities as a spa, Tom Weiskopf-designed golf course, and beach, encouraged by new construction that will never betray its age.

Top: Villa.

Above: Guest bedroom.

Right: Guest living room with high ceiling.

Opposite bottom left: Guest living room.

Opposite bottom left: Bar.

Photography: Kerzner International.

HKS Hill Glazier Studio

Venetian Macau Resort Hotel Casino
Macau, China

Inconceivable just 10 years ago, the 3,000-suite, 10.6 million-square-foot Venetian Macau Resort Hotel Casino has opened in Macau, China, reprising the Las Vegas-style Venice of the original Venetian Las Vegas—complete with replicas of St. Mark's Campanile, Doge's Palace and Clock Tower—on the Cotai Strip, a 200-acre expanse of reclaimed land. This trebling of the Venetian Las Vegas, designed by HKS Hill Glazier Studio with interior design by Wilson & Associates, surpasses the original in visual richness without losing spatial orientation. Its 39-story tower and five-story podium celebrate engineering and construction as well as design, rising from 14,000 precast concrete piles driven 200 feet deep, and assembling elaborate facades from a 361-piece "kit of parts." Guests typically focus on specific facilities, however, from a 550,000-square-foot Casino—the world's largest at 3,400 slot machines and 870 gaming tables—to 1.6 million square feet of retail stores, 15,000-seat Events Center, 1,800-seat Theater, 500,000-square-foot Congress, and 1.0 million-square-foot Expo Center. Superlatives aside, the Venetian Macau is thriving because it creates great guest experiences. "Follow the stately entry galleria," notes Jeffrey K. Jensen, AIA, hospitality project design director of HKS Hill Glazier Studio, "and you'll find yourself in the Casino—or a shopping center resembling the Grand Canal, complete with gondolas."

Top: Exterior.

Above middle left: Entry galleria.

Above middle right: Gondolier on canal in shopping center.

Left: Rotunda.

Photography: Visual Media Department, Venetian Macau Limited.

Intra-Spec Hospitality Design, Inc.

4218 Glencoe Avenue
Suite 1
Marina Del Rey, CA 90292
310.821.0376
310.821.7151 (Fax)
www.intra-spec.com

Intra-Spec Hospitality Design, Inc.

Hotel Solamar
San Diego, California

San Diego is known for its flawless Mediterranean climate, gracious parks and 70 miles of beaches as well as its telecommunications, biotechnology, software and electronics industries. Therefore, it attracts cool trendsetters along with retired servicemen from its former Naval Training Center. In fact, the traveler seeking hip luxury accommodations can encamp right in the historic Gaslamp Quarter at the new 235-room Hotel Solamar, a Kimpton Hotel with interior design by Intra-Spec Hospitality Design. The interior, featuring a lobby, meeting spaces, ballroom and business center in addition to guestrooms and suites, represents an energetic, sensual yet playful synthesis of San Diego's nautical heritage with contemporary design concepts. It creates its unique style by filling free flowing spaces with intimate, small-scaled areas appointed in exotic furnishings and overscaled furniture, and surrounds guests in an exuberant spectrum ranging from vibrant pinks and greens to chocolate browns, crisp whites and blues. Following a day at such nearby attractions as the Gaslamp Quarter, Petco Park, home of the Padres, San Diego Convention Center, San Diego Zoo and the museums of Balboa Park, the Solamar gives the "in" crowd a place to call home.

Top left: Living/dining room in suite.

Top right: Guest bath.

Above left: Guestroom.

Left: Meeting area prefunction.

Opposite top left: Lounge.

Opposite bottom: Lobby living room.

Photography: Glenn Cornier.

Intra-Spec Hospitality Design, Inc.

Lake Arrowhead Resort and Spa
Lake Arrowhead, California

Like finding the perfect picture frame to complement a painting, a major remodeling of Lake Arrowhead Resort and Spa, a 177-room resort hotel and spa in Lake Arrowhead, California, designed by Intra-Spec Hospitality Design, has revitalized the 25-year-old facility so that its lobby, boardroom and other meeting rooms, ballroom, pre-function space, corridors, guestrooms and suites are now a match for the San Bernardino Mountains that surround them. The hotel's public spaces, for example, have been enhanced by meeting rooms that can accommodate both large and small events, including wedding parties, and a main lobby that displays a massive stone fireplace previously obscured by a suspended walkway. Similarly, guestrooms now offer seating areas that frame views of guestroom fireplaces and Lake Arrowhead while retaining their intimacy. Throughout the hotel, the new, open arrangements of rooms and furnishings celebrate the rustic charm and sophisticated comforts of the hotel. Finely crafted natural and often indigenous materials such as mahogany, pine and teak, leather, Artesian glass and granite pay homage the mountain location. Comfortable, lodge-style furnishings and custom lighting fixtures, which mirror the shapes and contours of leaves, birds, and other natural forms, further reflect the hotel's connection to nature and remind guests that the hotel now offers as many internal delights as exist in the great outdoors beyond

Above left: Reception.

Top right: Spa journey suite.

Above middle: Pinnacle boardroom.

Above: Spa jacuzzi.

Opposite top left: Double guestroom.

Opposite top right: Corner suite.

Opposite bottom: Lobby fireplace.

Photography: Rouse Photography.

Intra-Spec Hospitality Design, Inc.

The Muse Hotel
New York, New York

When The Muse Hotel, a Kimpton Hotel, stepped into the limelight recently in New York's legendary Theater District, there was every reason to expect the interiors to capture the spirit of the performing arts. After all, this new, 200-room boutique hotel is steps away from Broadway, Times Square and landmark theaters, attracting theatergoers as well as adventurous business people. Sure enough, the sleek design draws inspiration from the nine Greek muses, and embraces a Grecian theme in its conversion of the historic Leavitt Building. Key public areas such as the corridors, meeting rooms and Wine Cellar private dining room have been designed by Intra-Spec Hospitality Design, along with 200 guestrooms and 5 suites. They are a sophisticated, upscale oasis, combining ebony wood flooring, silken fabric-covered walls, polished chrome and clear crystal with sleek lacquered case goods and dramatically styled seating to impart sparkle and glamour to the guest experience. A prized feature is that guestrooms, equipped with such amenities as super-soft feather beds, WiFi Internet service, flat-screen TVs, and DVD/CD players, are large—some 50 percent larger than those in most New York hotels—something most New Yorkers only dream about.

Top right: Bed showing headboard and linen.

Below right: Lounge chair.

Bottom right: Bedside table and lounge chair.

Bottom: Guestroom chest and Muse bust.

Opposite: Guestroom.

Photography: David Phelps.

Intra-Spec Hospitality Design, Inc.

Mandarin Oriental
San Francisco, California

Awe-inspiring views from the guestroom windows of the Mandarin Oriental in San Francisco are among the numerous reasons why guests favor this 158-room luxury hotel, located within a high-rise tower in the Financial District. As befits a leading name in hospitality, the Mandarin Oriental pampers San Francisco guests with meticulous service and such elegant facilities as its reception/lobby, concierge desk, restaurant, bar, business center and fitness center. Nevertheless, the guestrooms and suites are what many guests remember and describe. To enhance their appeal, the Mandarin Oriental invited Intra-Spec Hospitality Design to update their interior appointments and maximize their views, a project covering some 80,000 square feet on floors 38 through 48. The design team's strategy involved raising beds high enough to make views visible even for reclining guests, adding Roman sheer window coverings to diffuse light and aid views, and freshening overall décor with clean, modern furnishings, drawing on mahogany, gold leaf, black granite, bamboo, carpet and textured silken fabrics as well as a color scheme of rich reds and glowing golds to blend Asian motifs with contemporary design themes. Has the fine-tuning paid off? Trip Advisor observes, "The bedrooms are essentially like jewel boxes in the sky...."

Top right: Guestroom.

Middle right: Guest suite.

Right: Guest suite at evening.

Photography: George Apostotlides.

JCJ Architecture

Hartford
38 Prospect Street
Hartford, CT 06103
860.247.9226
860.524.8067 (Fax)
www.jcj.com

New York
Empire State Building, Suite 1029
350 Fifth Avenue
New York, NY 10118
212.774.3606
212.774.3607 (Fax)

San Diego
3838 Camino Del Rio North
Suite 361
San Diego, CA 92108
619.282.9922
619.282.6678 (Fax)

Phoenix
2141 East Highland Avenue
Phoenix, AZ 85016
602.957.5060
602.954.4251 (Fax)

JCJ Architecture

Seneca Niagara Casino & Hotel
Niagara Territory, Niagara Falls, New York

For years, honeymoon couples and other visitors have admired magnificent Niagara Falls from the American side and then patronized tourist-oriented attractions on the Canadian side, ignoring Niagara Falls, New York, 20 miles north of Buffalo. But the Seneca Nation of Indians have found a solution through the Seneca Niagara Casino & Hotel, which features architecture and interior design by JCJ Architecture. The 147,000-square-foot Casino and 604-room, 26-story, 731,000-square-foot Hotel have forever and dramatically altered the aging industrial city's dynamics by offering visitors unprecedented new entertainment options. Now choices range from gaming activites —4,289 slot machines and 115 table and poker games—to a spa, salon, fitness area and swimming pool, six restaurants, 25,000-square-foot multi-purpose entertainment facility, 468-seat theater, 8,000 square feet of conference and banquet space, five retail stores, and parking facilities for over 3,700 vehicles.
A catalyst for the Seneca Nation of Indians' "Vision for the Future Master Plan" to restore the city's glory and secure tribal sovereignty and economic self-determination, the award-winning project is highly effective at projecting its bold, contemporary imagery, while integrating subtle references to tribal themes and iconography, in large and small scale settings. The Casino, for example, has transformed Niagara Falls' former convention center (designed by Philip Johnson in 1974) by reviving its soaring, open space with an internally lit, three-dimensional Lexan sculptural screen in place of the original curtain wall, a vibrant gaming core of slot machines and table games, and a superstructure of sculptural steel and colorful, sail-like canvas canopies that humanize the main floor's

Top: High limit slot area.

Right: Exterior of casino.

Far right: Gaming floor.

Opposite bottom left: Hotel lobby.

Opposite bottom right: High limit gaming area.

Photography: Steve Lakatos Photography, Robert Benson Photography, Ian Vaughan Productions Inc.

RESTAURANTS
LOGO SHOP
PROMOTIONS
GARAGE
GUEST SERVICES
CAFE
RETAIL
CASINO
RECEPTION
VALET
BELL STAND
VIP

JCJ Architecture

Left: The Western Door: A Seneca Steakhouse.

Below far left: Showroom.

Below left: La Cascata Restaurant.

Bottom: Corner guest suite.

Opposite: La Cascata Restaurant.

scale and accommodate specialty lighting and surveillance equipment overhead. Rising behind the Casino, the new, high-rise Hotel displays an abstract depiction of mountains, sky and historic landscape on a reflectively sheathed façade, incorporating such references to the Seneca Nation as a signature "single feather" adorning the top, 300 foot tall stripes of multi-colored lighting evoking falling water, and abstract woodlands, all atop a stone base embodying the land. Not surprisingly, the interior is equally eloquent. In the lobby, the decorative, illuminated frieze panels above the entry doors recall the "rainbow over the mountains" motif from the Territory in Salamanca, while dramatic waterwalls flanking the Casino entry honor the highly-respected and protected water sources running through the tribe's woodlands territory. What do guests from New York, Pennsylvania and Ohio think of Niagara Falls today? Some 30,000 visit the Seneca Nation's complex daily, and rooms are fully booked.

JCJ Architecture

Seneca Allegany Casino & Hotel
Allegany Territory, Salamanca, New York

Left: Exterior showing main entrance.

Right: Patria Restaurant.

Below left: Turtle Island non-smoking casino.

Below right: The Western Door: A Seneca Steakhouse.

Photography: Ian Vaughan Productions, Inc.

As people of the woodlands along southwestern New York's Southern Tier, the Seneca Nation of Indians was determined to see the new Seneca Allegany Casino & Hotel, in Salamanca, New York, reflect its ancestral home and cultural heritage, as well as establish a spectacular entertainment venue. JCJ Architecture has convincingly achieved the goals of the Seneca Nation for the 345,000-square-foot facility, including a two-level, 68,000-square-foot Casino and 212-room, 11-story Hotel. The results can be seen and admired in a variety of public spaces such as the gaming facility, containing 2,235 slot machines and 56 table and poker games, casino bar, five themed restaurants, guestrooms and suites, world-class spa, workout facility and indoor pool, and 3,000-square-foot retail shop. To make the project a reality, the design simultaneously acknowledges key spatial relationships required for effective casino and hotel operations, prevailing site conditions with dramatically steep terrain, and the aspirations of tribal members to re-interpret cultural themes and select Seneca artifacts – a joint effort between JCJ Architecture, which has served the Seneca Nation since 2000, and the Nation's Cultural Enhancement Resource Committee, Design Review Committee, Tribal Council, and Gaming Corporation Board. Consequently, the resort that greets patrons is both visually compelling and genuinely meaningful. On the exterior, the hotel tower's vivid blue reflective glass and simple yet bold

Left: Casino bar.

Below: Fireplace seating in lobby.

Below middle: Center Suite.

Bottom: Spa swimming pool.

form, resting on a robust base of elemental forms sheathed in white metal panels and synthetic stucco, evokes three natural affinities of the Seneca: sky, water, and woodlands. Within the interior, finishes and detailing identify and differentiate the lower level, a "daytime" setting housing gaming facilities and connections to the adjacent hotel and spa, apart from the upper level, where food and beverage establishments revel in a "nighttime" milieu. Not only is the resort busy and vibrant 24 hours a day, 7 days a week, the Seneca Nation now has a vital new symbol. Notes Maurice John, Sr., president of the Seneca Nation of Indians, Bear Clan, "This glittering jewel in the country reflects the natural beauty of its environment and our proud heritage."

Klai Juba Architects

4444 West Russell Road
Suite J
Las Vegas, NV 89118
702.221.2254
702.221.7655 (Fax)
www.klaijuba.com

Klai Juba Architects

Seminole Hard Rock Hotel and Casino
Hollywood, Florida

With Miami Beach's sizzling South Beach just steps away, the new Seminole Hard Rock Hotel and Casino in Hollywood, Florida, has diligently transformed a 100-acre site into a miniature Las Vegas with gaming, hotel, dining and entertainment venues delivering the "Hard Rock" theme. Does the project meet the demands of south Florida's dynamic travel and leisure market? Travel writers and customers say the Hard Rock has hit the jackpot with a 481-room, 870,000 square-foot complex that includes a 130,000 square-foot casino, 17 restaurants and lounges, meeting/convention facilities, shopping, an entertainment complex, a 22,000 square-foot spa, and a 4.5-acre lagoon style outdoor pool for which Klai Juba Architects provided comprehensive property planning and design, while at the same time serving as the executive architect. Key to the Hard Rock's success has been the seamless integration of top-notch front-of-house spaces with efficient back-of-house planning. Of course, customers have also noticed that guest accommodations are surprisingly stylish and comfortable; featuring contemporary furnishings and spacious bathrooms. Due to the success of the Hard Rock in Hollywood and a sister Hard Rock in Tampa, Florida, the Seminole Tribe is continuing to use Klai Juba Architects to help plan future developments for both properties - truly a win-win game plan.

Top: Exterior with parking garage.

Above: Registration / Lobby.

Left: Evening view of main façade.

Opposite: Poolside accommodations.

Photography: Seminole Hard Rock Hotel and Casino.

Klai Juba Architects

Seminole Hard Rock Hotel and Casino
Tampa, Florida

Below: Pool entrance.
Below right: Evening view.
Opposite: Center bar.
Photography: Seminole Hard Rock Hotel and Casino.

How do you bring the Hard Rock brand of gaming, hotel, dining and entertainment facilities to Tampa, Florida? For the Seminole Tribe and Klai Juba Architects (providing comprehensive property planning and design, while at the same time serving as executive architect), the winning formula has been its appeal to a diversity of customers, welcoming adult leisure and business travelers as well as young families with children. The 12 story, 400,000 square-foot Art Deco-style complex includes 250 guestrooms and suites, a 90,000 square-foot casino, restaurants, bars, a ballroom, a meeting facility, entertainment venues, a spa/ health club and an outdoor pool - all strategically located adjacent to the Tampa amphitheatre and minutes from Busch Gardens, the Florida Aquarium, and downtown Tampa. More importantly,

the Hard Rock environment is effective, versatile and appealing. The 8,000 square-foot ballroom can be divided into eight sections, the spa/health club offers many luxurious amenities and spa services, and the modern pool is equipped with cascading fountains and poolside cabanas. The well appointed guestrooms boast sunny views and even naturally lighted bathrooms. The Seminole Tribe now has Klai Juba Architects planning even new ways to keep Tampa Hard Rocking.

Klai Juba Architects

Mandalay Bay Resort and Casino
Las Vegas, Nevada

It's a long road from Mandalay, the second largest city in Myanmar (formerly Burma), to the Mandalay Bay Resort and Casino in Las Vegas. Fortunately, guests who visit the 4,332-room, twelve million-plus-square-foot complex (master planned and designed by Klai Juba Architects), find themselves instantly transported to its celebrated South Seas setting, and have been enthusiastically doing so since opening day. In fact, business was so strong that Klai Juba Architects continued to be the architect of record for all property development from Mandalay Bay's opening in 1999 to its sale to MGM Mirage in 2005. The full-service hotel combines several resort destinations including an integrated Four Seasons Hotel and boutique hotel experience named THEhotel, a 135,000-square-foot casino, and a 2 million-square-foot convention center. Also featured is the aquarium, wedding chapel, event center, theater, three spas, numerous restaurants and nightclubs. The tropical pool area includes a wave pool, beach club and the House of Blues concert venue, and who can forget the impressive Mandalay Place - a 150,000 square-foot retail venue on a sky bridge connecting the hotel to the neighboring Luxor and Excalibur Hotel and Casino, completing Mandalay's vision of encompassing 13,000 rooms under one roof. Comprehensive planning and design have produced a single cohesive property whose vast front-of-house and back-of-house operations are so smoothly integrated that guests can focus on just one thing... having a great time.

Above: View of exterior through archway.

Below: Mandalay Bay with THEhotel.

Opposite: Exterior.

Photography: Courtesy of Klai Juba Architects.

MANDALAY BAY
MANDALAY BAY

Klai Juba Architects

Hard Rock Hotel and Casino
Las Vegas, Nevada

Fodor's describes the Hard Rock Hotel and Casino in Las Vegas, Nevada in words few Sin City visitors need to read from a guide book: "The Hard Rock doesn't try to be cool - it just is - and it is the 'cool-without-trying' who hang out here." The original establishment opened relatively modestly in size (300 rooms) in 1995 and nearly doubled in size in 1999, reflecting the market's swift acceptance of the brand and the concept. Now Klai Juba Architects has been retained to design the nearly 2 million square-foot addition, including a 200,000 square-foot low-rise structure, two towers of 900 guestrooms and suites, and an additional parking garage. The expansion will offer guests the exciting experience people expect from Hard Rock on a much larger scale. Upon completion the casino will expand from 30,000 square feet to 65,000 square feet, with additional food and beverage outlets along with retail and meeting space expanding significantly. Expansion plans also include "The Joint", a 4,000 seat intimate, yet full service, concert venue. The Hard Rock Hotel and Casino, like other major gaming hotels, dining and entertainment venues in Las Vegas, agrees with Mae West that "Too much of a good thing is wonderful."

Above: New hotel tower.
Below: Aerial view of site.
Illustration: Courtesy of Klai Juba Architects.

Knauer Incorporated

720 Waukegan Road
Suite 200
Deerfield, IL 60015
847.948.9500
847.948.9599 (Fax)
www.knauerinc.com

Knauer Incorporated

Entourage
Schaumburg, Illinois

Right: Exterior.
Below: Wine room.
Below middle: Bar.
Bottom: Main dining room.
Opposite: Lobby.
Photography: Mark Ballogg.

Frank Sinatra's Rat Pack would have loved the big martinis, blue cheese-crusted steaks, high-style Retro look in stone, wood and glass, and anything-goes ambiance at Entourage. This new, 775-seat, two-story, 22,000-square-foot steakhouse on a five-acre site in Schaumburg, Illinois was designed by Knauer Incorporated. The concept behind Entourage, providing a place to pamper guests and their "entourages" with comfort, congeniality and service, has been vividly realized. This is a freestanding structure that updates Atomic Age architecture with a striking, futuristic exterior—dramatized by a 48-foot-high, illuminated "martini shaker" glass window—and a crisply tailored, club-style interior that guests enter through a lobby featuring an open, winding staircase and dual 40-foot-high waterfalls. It's not by accident that the lobby, bar, main dining room, wine room, wine cellar, banquet room and exhibition kitchen seem "shaken, not stirred." Playing rusticated field stone piers and walls against precisely milled cherry wood columns, wainscoting and floating soffits, generous banquettes and other clean, contemporary furnishings, two fireplaces, and dramatic lighting, the design establishes a suave and sumptuous setting that portrays every guest as a VIP. If you don't see Ole Blue Eyes, Dino or Sammy Davis Jr. here, you and your "entourage" can play their roles yourselves.

Knauer Incorporated

InterContinental Milwaukee
Milwaukee, Wisconsin

Everything says cool, hip and classy. Guests of the former Wyndham Milwaukee Center surely never regarded their nondescript hotel in such terms, but that's how young professionals are describing the new InterContinental Milwaukee, a popular, 221-room hotel whose comprehensive renovation was designed by Knauer Incorporated. The transformation stems from Knauer's decision to reposition the hotel in the marketplace, devising a new, chic and trendsetting DNA to redefine every physical aspect of the hotel product, from the lobby, lobby bar/café, restaurant, kitchen, ballroom and health club to guestrooms and suites. For a hotel in the center of Milwaukee's growing theater district (the Marcus Center for the Performing Arts stands across the street design also by Knauer Incorporated), the strategy has produced facilities that perfectly match the neighborhood. Kil@wat restaurant, for example, mixes long, horizontal soffits and wall cut-outs with cubes of light, illuminated scrim curtains, red walls and wood tabletops to create a study in perpetual motion. By contrast, the typical guestroom is appointed as an urban pied-a-terre offering simple, artful and quiet comfort, where the bedroom resembles a living room, and the bathroom resembles a spa. Despite the renovation's modest cost, the InterContinental now leads Milwaukee in occupancy and room rates. Cool.

Top left: Lobby lounge.

Top right and upper right: Guestroom bedroom and bath.

Above right and right: Kil@wat restaurant.

Opposite: Reception.

Photography: Mark Ballogg.

KILAWAT
illuminating cuisine

Knauer Incorporated

Cameron's Steakhouse
Birmingham, Michigan

America's love affair with steak burns with unabated flame, so a new steakhouse that turns heads as well as satisfies carnivores gets attention even within the crowded, high-end restaurant scene of greater Detroit, where 275-seat, 8,000-square-foot Cameron's Steakhouse is drawing crowds to suburban Birmingham. To help put Cameron's on the gastronomic map, Knauer Incorporated has designed a new paradigm for the suburban steakhouse: an elegant, contemporary environment with the cool sophistication of downtown—and none of the timeworn "men's club" clichés. Cameron's is unapologetic about putting guests in a formal, upscale setting of sweeping architectural lines, neutral tones spiked with bright accents, smooth surfaces of wood, stone and glass, cleanly detailed "power booths" and other contemporary furnishings, jazz-influenced artwork, and warm lighting. Thus, the steaks look good and the guests look even better, assuring that the national passion for filet mignon and porterhouse will have another great venue to perpetuate it.

Above: Dining room.
Left: Bar.
Photography: Mark Ballogg.

Knauer Incorporated

Lux Bar
Chicago, Illinois

Does good, everyday food taste better served in a classic Modernist environment? Lux Bar, Chicago's recently completed architectural homage to Viennese Modernist Adolf Loos, recreating his Bar Americane (called the Loos Bar by today's Viennese), has succeeded admirably in doing this. Guests patronizing Lux Bar's 300 plus-seat, three-story, 10,500-square-foot facility, designed by Knauer Incorporated, may not even notice how efficiently the scheme uses space at one of the Windy City's busiest restaurant corners, to fit the kitchen in the basement, the bar and restaurant on the first floor, and the banquet space on the second floor. Instead, everyone is mesmerized by Lux Bar's black-and-white interiors, which employ mahogany, glass, mirrors, brass and marble, custom furnishings and custom lighting fixtures to recreate the Viennese Secession look. Though nostalgia doesn't come cheaply, the finished product is a stunning backdrop for such entrées as buttermilk-fried chicken, "Gold Coast" sliders, and eggs all the time.

Top: Dining room.
Above left: Exterior.
Above left: Bar.
Photography: Mark Ballogg.

Knauer Incorporated

Bacchus
Milwaukee, Wisconsin

Right: Bar.
Below: Main dining room.
Photography: Mark Ballogg.

Every city needs an appropriate place for power lunches, corporate or social dinners, and beautiful wedding receptions, which explains the recent opening of 250-seat, 5,500-square-foot Bacchus, designed by Knauer Incorporated, in downtown Milwaukee's historic Cudahy Tower. This decidedly upscale restaurant and bar, showcasing fresh seafood, handmade pastas, and grilled meats, has been developed to establish a new standard for fine dining that embraces décor as well as service. True to its strategy, Bacchus provides a variety of event spaces, including private dining rooms and an outdoor patio, to supplement a main dining room and bar, all designed and constructed with the beauty, precision and craftsmanship of exquisite leather goods. In fact, the classic contemporary environment actually incorporates leather walls and details along with wood and glass, handsome contemporary furnishings, and a variety of architectural and decorative lighting fixtures to achieve its unusually high level of finish. Milwaukee should have no difficulty finding good occasions for visiting Bacchus.

LBL Architecture & Interiors

235 Ocean Park Blvd.
Suite B
Santa Monica, CA 90405
310.450.8900
310.450.8962 (Fax)
www.lblarc.com

LBL Architecture & Interiors

Oak Valley Ski Resort
Wonju City, Korea

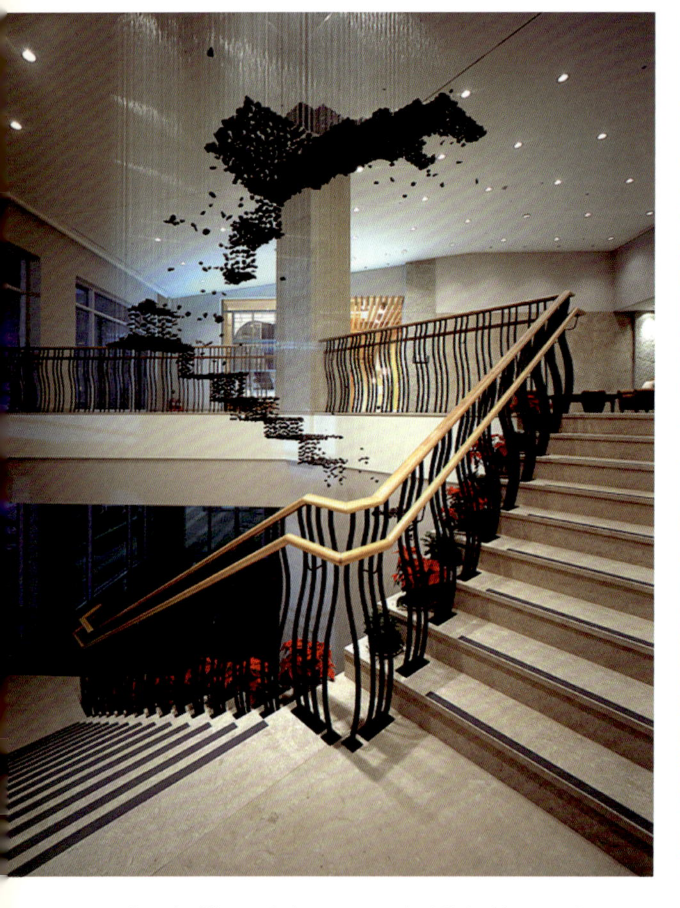

Superb skiing at Oak Valley Ski Resort, in Wonju City, South Korea, is what sports enthusiasts expect to find across Gangwon province, located in the Korean peninsula's central eastern region. Gangwon-do offers tourists a ruggedly beautiful terrain where some 82 percent of the land is mountainous and snowfall is plentiful. But visitors are discovering that Oak Valley, a property of Hansol Development Co., Ltd. that features space planning and interior design by LBL Architecture & Interiors, offers the perfect setting for another sport once skiing season ends: golf. Adapting the sloping grounds for serious golfers, a project undertaken with an experienced golf course consultant, was just one challenge among many. However, the 250-room resort, whose light, naturally finished contemporary interiors also include a lobby/lounge, restaurant, banquet rooms, spa and ski/golf clubhouse, convincingly establishes the right mood for every season. Its design, blending such timeless materials as limestone, granite, slate, maple wood, glass mosaic tile and art glass, balances elements of warmth and coolness to complement its two principal sports activities. Highlighted by such details as finely crafted millwork, stylish furniture, distinctive staircases, handsome textiles and carpet, and a sophisticated lighting scheme, Oak Valley Ski Resort should inspire fond memories year-round.

Top left: Staircase to lobby.

Top right: Restaurant booth seating.

Above: Café table seating in restaurant.

Opposite top: Lobby.

Opposite bottom left: Powder room.

Opposite bottom middle: Spa.

Opposite bottom right: Exterior.

Photography: Courtesy of Kesson International.

LBL Architecture & Interiors

Hokusai
Beverly Hills, California

Behind an Art Deco façade at the busy intersection of Wilshire Boulevard and Gale Drive in Beverly Hills, Hokusai, a Japanese sushi bar and restaurant, has just performed a small miracle of transformation. A once dark and musty French restaurant now thrives as a modern and smartly tailored showcase for Euro-Japanese Fusion cuisine. Named for the great 19th-century ukiyo-e master and designed by LBL Architecture & Interiors, the 87-seat, 3,000-square foot establishment could easily fit into the trendy nighttime scene in Tokyo's Ginza or Shibuya districts. Its space has been reconfigured to include a dining room, sushi bar, bar/lounge and kitchen by relocating the existing bar from the center of the space to one side and shrinking the existing kitchen just enough to introduce a sushi bar with a view of the kitchen. To create its simple, dark and stylish persona, the design blends limestone, stainless steel and zebrawood with contemporary furnishings, understated lighting and photographs of seasonal foliage, producing a series of intimate environments. If the enthusiastic young customers coming to Hokusai are any indication, the new restaurant is serving the perfect milieu for such distinctive entrées as Saikyo foie gras, Jidori chicken and Kobe beef cheeks.

Top right: Entrance to dining room.

Above middle: Sushi bar.

Above: Booth seating.

Left: Bar and banquette seating.

Opposite: Dining room and sushi bar.

Photography: Courtesy of Dennery Marks.

LBL Architecture & Interiors

Oakwood Premier COEX Center
Penthouse Presidential Suites
Seoul, Korea

Business travelers who demand the ultimate in accommodations when visiting Seoul can now enjoy the comfort of a private home with the service of a five-star luxury hotel at Oakwood Premier COEX Center. Oakwood Worldwide, a global operator of corporate housing, temporary housing and serviced apartments, has spared no expense to develop two sumptuous Penthouse Presidential Suites at Oakwood Premier COEX Center, a 280-unit serviced residence at the Korean World Trade Center complex in Seoul's central business district, Gangnam-gu. Each of the 03-bedroom, 4,320-square foot Presidential Suites, designed by LBL Architecture & Interiors in Neo-classical style, has separate living, dining, and sleeping areas in addition to a gourmet kitchen, wet bar and laundry. Themed with "Gold" or "Red" accents, the residences are appointed in such timeless materials as marble, cherry wood and Venetian plaster, along with custom cabinetry and millwork, period-style furniture, sumptuous textiles, rugs and carpet, and exquisitely detailed decorative lighting. With daily maid service, 24-hour guest relations, security, concierge and room service, and such house amenities as a residents' lounge, restaurant and bar, business center, several other on-site restaurants, and health club, guests of the Presidential Suites will feel like presidents regardless of their actual business titles.

Top left: Bedroom.
Top right: Entrance lobby.
Above right: Entry foyer.
Above middle: Dressing room.
Above left: Kitchen.
Far left: Jacuzzi bath.
Left: Wet bar.
Photography: Taeho Jeong.

소화기

LBL Architecture & Interiors

COEX Seven Luck Casino
Seoul, Korea

Like a city within a city, the COEX Convention & Exhibition Center, South Korea's largest convention center, draws up to 200,000 people daily to its location at the Korean World Trade Center complex in Seoul's Gangnam district. Boasting state-of-the-art business facilities, COEX also hosts numerous diversions, including the COEX shopping mall, two luxury hotels, multiplex cinema, aquarium—and now the nation's largest "foreigners-only" casino, COEX Seven Luck Casino, designed by LBL Architecture & Interiors. A multi-story, 50,000-square-foot facility, COEX Seven Luck makes the most of its narrow, "shot-gun" space by incorporating an entry reception area on the first floor, casino game rooms, lounges, restaurant and kitchen on the second and third floors, and spa and fitness center on the fourth and fifth floors. Customers are encouraged to explore it through a bold, contemporary design where sophisticated lighting dramatizes a geometric architectural interior of granite, marble, bronze, hardwoods, carpet and glass tile. While international in spirit, the casino subtly evokes Korean culture through such motifs as the game room doors, restaurant décor, and traditional tiled roof ("katt") and lanterns above each gaming table in the casino rooms. Its exciting ambiance demonstrates that 400,000-square foot COEX can do small things very well.

Top left: Money Bar.

Top right: Entrance façade.

Middle right: Railing detail in atrium.

Above: Casino.

Left: Atrium.

Photography: Charles Lee.

Mancini•Duffy

39 West 13th Street
New York, NY 10011
212.938.1260
800.298.0868
212.938.1267 (Fax)

New York
New Jersey
Connecticut
Washington, DC
London, UK

www.manciniduffy.com
info@manciniduffy.com

Mancini•Duffy

Equinox Fitness Clubs
Multiple Locations

Exercise, like spinach, is good for you—though Mother never promised you'd enjoy it. How does Equinox Fitness Clubs, an operator of upscale, full-service fitness clubs founded in New York in 1991, make exercise enjoyable and even glamorous for club members in greater New York, Los Angeles, San Francisco and Chicago? Key to its success has been the use of edgy, contemporary and distinctive design to support a customer experience built on specialized programs and services provided by highly qualified trainers. As part of an on-call agreement with Equinox, Mancini•Duffy has provided architectural and interior design services for clubs in several Manhattan and Long Island locations as well as in Chicago, combining the Equinox brand with unique interpretations of local cultural and demographic themes. Every facility maintains the company's high standards of design and construction in establishing clubs that have at least 25,000 square feet of studios devoted to such activities as aerobics, yoga, pilates, spinning and dance, retail functions such as cafés/juice bars, sports apparel shops, and beauty/spa product shops, and such additional services as fitness and nutrition, day care/baby sitting, spa, pool, sauna, steam, and cardio and strength training. Even mother would enjoy it. In fact, many mothers probably do.

Top left: Storefront entrance.

Top right: Pool.

Above: Pilates.

Right: Entry.

Opposite bottom left: Cardio and strength training.

Opposite bottom middle: Massage therapy.

Opposite bottom right: Restroom.

Photography: Wanye Cable Photography.

Mancini•Duffy

Rosa Mexicano
Palm Beach Gardens, Florida

Restaurateur Josefina Howard raised New Yorkers' expectations for Mexican cooking by introducing what *New York* magazine called in 1984 "a hitherto unfamiliar, elevated version of Mexican cuisine." Now, Rosa Mexicano has brought its distinctive cuisine and jubilant atmosphere to a new, 265-seat, 11,000-square-foot restaurant, designed by Mancini•Duffy, in Palm Beach Gardens, Florida. The fifth Rosa Mexicano transforms an irregularly shaped space with a 23-foot ceiling, extensive glazing, and loading dock into a festive and largely open front of house, where the bar/lounge provides a lively background for the main dining room, private dining room and outdoor dining patio, and an efficient back of house, enclosing the kitchen and other service areas. Expanding on the palette developed with the Rockwell Group for Rosa's Union Square location in New York, Mancini•Duffy has employed desert-inspired colors, dropped soffits, floating planes and faux wood ceiling beams with programmable LED lighting, simple and comfortable furnishings, lava stone, terrazzo, plaster walls, wood flooring, handmade tile, and Rosa's signature, internally-lighted blue glass tile water wall, positioned on axis with the entry to divide the bar from dining areas to create the perfect setting for dishes like Guacamole en Molcajete (guacamole made in a lava-rock mortar).

Top left: Illuminated wall detail.

Top right: Restroom.

Above left: Exterior.

Middle left: Bar.

Left: View from bar to main dining room.

Opposite top: Glass tile water wall.

Opposite: Main dining room.

Photography: Peter Paige Photography.

Mancini•Duffy

Bliss
New York, New York

Complementing a captivating menu of services with a fun, "no-attitude" atmosphere, Canadian fitness instructor Marcia Kilgore quickly attracted a headline clientele to her first Bliss spa in New York's trendy SoHo in 1996, leading her to successful skin care and home spa products. Bliss World is now part of Starwood Hotels & Resorts Worldwide, and Mancini•Duffy is providing design services for the roll-out bringing Bliss to Starwood's innovative W Hotels around the world. Each roll-out adapts the Bliss logo, cloud motif and high design standards to different site conditions. Consider a new, 5,125-square-foot New York facility, where Bliss offers a reception area, retail shop, treatment rooms (including one for wet treatment), lounges, manicure/pedicure, steam and showers, sauna, locker rooms and restrooms. Customers easily navigate its luxurious interiors, proceeding from reception and retail areas to increasingly private spaces, because the compact design extracts maximum usable space from the existing structure, arranges spaces to assure privacy, and promotes serenity and well-being by

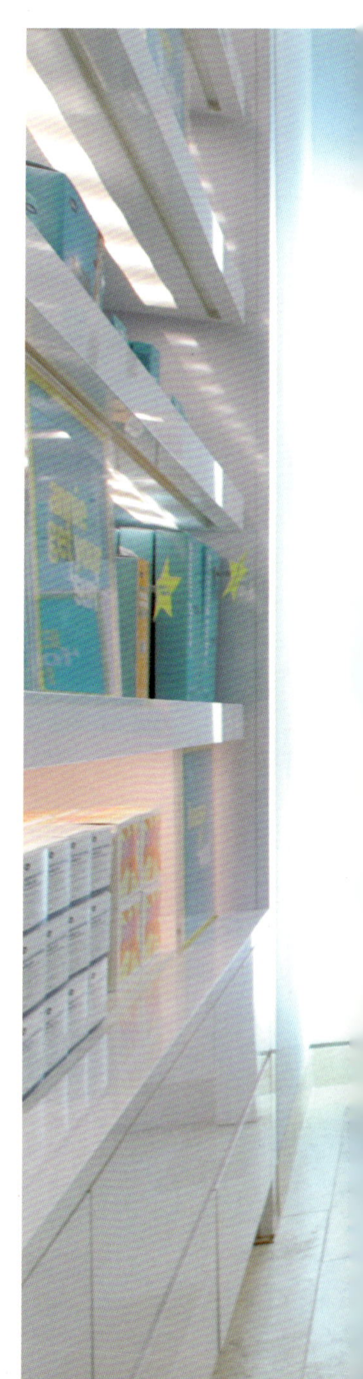

skillfully applying "Bliss blue" and warm cream and limestone colors, such classic materials as mosaic tile, art glass and exotic woods, informal, modern furnishings and soft lighting to raw space. Isn't that what should be expected of a space called Bliss?

Right: Corridor.

Below: Retail shop.

Opposite left: Shower.

Opposite right: Lounge.

Opposite middle left: Reception.

Opposite bottom left: Lounge.

Opposite bottom right: Restroom.

Photography: Robert Mitra.

Mancini•Duffy

Vento Trattoria
New York, New York

Who would suppose a derelict, former factory building constructed in 1849 and converted into a Civil War infirmary—before housing one marginal business after another—would become a stylish Italian restaurant where trendy guests order entrées like Pepperoni Imbottiti stuffed with baby sweet red peppers and parmigiano cheese? If see-and-be-seen dining is believing, the150-seat two-story Vento Trattoria, designed by Mancini•Duffy for restaurateur/hotelier Steve Hanson of B.R. Guest Resturants, has given its landmark structure in Manhattan's historic Meatpacking District a new lease on life. Vento's design, which encompasses a bar, dining rooms and kitchen, as well as a sidewalk café and a separate basement lounge (doing business as Level V), establishes an airy, modern environment while restoring its historic context. Key to the design has been returning the wedge-shaped structure to its original condition and inserting understated new interior elements. By stripping paint from brick walls, repairing the sheet metal frieze, opening up bricked-over windows and doorways, installing new hardwood floors, cleaning original cast-iron columns, and installing such attractive, contemporary additions as the bar, stair, furnishings and other finishing touches, created in collaboration with Meyer Davis Studio, Vento has earned Zagat's praise as a "place to be seen."

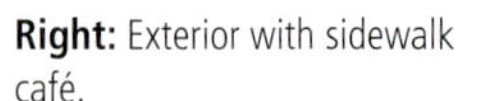

Right: Exterior with sidewalk café.

Below middle left: Stair detail.

Below middle right: Upper level dining room.

Below left: Street level dining room.

Bottom right: Bar.

Photography: Mancini•Duffy.

Marnell Architecture

222 Via Marnell Way
Las Vegas, NV 89119
702.739.2000
702.739.2045 (Fax)
www.marnellcorrao.com

Marnell Architecture

Riche Restaurant
Harrah's New Orleans Hotel
New Orleans, Louisiana

Fabulous cuisine honoring the French, Spanish, Cajun and Creole traditions is one of New Orleans's legendary gifts to the world. So the recent arrival of prominent Boston chef Todd English to open a classic French brasserie, Riche, at the new Harrah's New Orleans is probably not what the Crescent City's residents or visitors expected. Then again, the debut of Harrah's New Orleans in the historic Warehouse District—the city's first new hotel since Hurricane Katrina's rampage—has come as a pleasant surprise as well. Riche, a 166-seat, 6,000-square-foot restaurant designed by Marnell Architecture, continues the trend among hotel restaurants by establishing an independent identity for its facilities, which include a bar, dining room, open grill for exhibition cooking, oyster bar, 70-seat outdoor terrace and after-hours club, that should appeal to a broader clientele than hotel guests alone. While taking inspiration from its 19th century neighborhood, Riche adds idiosyncratic, contemporary touches of its own to create an open and bustling environment with pockets of privacy for those who want it. Together with 528, the jazz club English has established next door on Fulton Street in an historic 1852 building, Riche joyously announces that "The Big Easy" is ready to party again.

Top: Outdoor terrace.

Above: Bar.

Opposite: View of bar from entrance.

Photography: Insite Architectural Photography.

Marnell Architecture

Cirque du Soleil
The Beatles "Love" Cirque du Soleil Theatre
Las Vegas, Nevada

Love? Beatles fans are instantly assured that it's all you need when music fills the dazzling new, 2,012-seat Beatles "Love" Cirque du Soleil Theatre, designed by Marnell Architecture as architect of record and interior designer, at the Mirage Hotel & Casino in Las Vegas. Working from a suggestion by the late George Harrison, The Beatles, Apple Records and Cirque du Soleil have created "Love," a musical production in which 60 Cirque artists interpret the Fab Four's repertoire. The space once occupied by Siegfried & Roy now houses "Love" in a dynamic in-the-round theater supported by a box office, retail boutique, entry lobby and concession area. While much of the design focuses on accommodating the Cirque's cast, audio/visual equipment, set and props, audience members are scarcely neglected by a sleek, high-tech design that uses terrazzo floors, chrome walls and ceilings and a massive array of LED, fiber optic and custom lighting fixtures to draw them through a series of visual "adventures" to the main show. Cool and inviting as the public spaces are, they become irresistible when a preset, automated lighting design springs to life, suggesting that maybe "Love" is really all you need.

Above left: Retail boutique.
Above right: Box office.
Opposite bottom: Entry lobby.
Photography: Opulence Studios.

THE
BEATLES
CIRQUE DU SOLEIL
LOVE
BOX OFFICE

Marnell Architecture

Harrah's New Orleans Hotel
New Orleans, Louisiana

Even as New Orleans continues to recover from Hurricane Katrina, its lights shine noticeably brighter with the arrival of Harrah's New Orleans, the first new hotel built since September 2005. The 450-room, 450,000-square-foot Harrah's, designed by Marnell Architecture as master planner, executive design architect and interior designer, brings a contemporary boutique hotel to the historic Warehouse District, across the street from Harrah's existing Casino, on the edge of the fabled French Quarter. The new construction encompasses a porte cochere, lobby, over 7,000 square feet of meeting rooms, a fine dining outlet, jazz club, exercise facility, guestrooms and suites. To keep from overwhelming its 19th-century neighbors, the hotel is articulated as a traditional, low-rise base structure, including public areas and amenities, supporting a modern, 25-story tower, where guestrooms and suites overlook the Crescent City. The hotel transcends mimicry to harmonize with its surroundings. Working diligently with the New Orleans Historic District Landmarks Commission, the design team

Top: Suite living room.

Above: Suite master bath.

Opposite: Entry foyer of suite.

Photography: Insite Architectural Photography.

COLONY

Marnell Architecture

has faithfully replicated the 19th-century facades of the area's original buildings at the base—even incorporating salvaged materials from them—and has installed a jazz club within an adjacent 1852 building, a restored landmark registered with the National Trust for Historic Preservation, that is now part of the hotel. Inside, the elegant transitional environment of stone, wood, plaster, plush carpet and sumptuous textiles gracefully blends European and American furnishings with artwork commissioned from local artists to remind guests that Harrah's New Orleans welcomes them to a singular European city in 21st-century America. Guests who appreciate distinctive features will be pleased to find such delights as celebrity chef Todd English's Riche Restaurant, his interpretation of a French brasserie on Fulton Street, the all-suite 26th floor with butler service, and the lobby, a showcase for a stunning glass wall sculpture from artist Gene Koss. Wherever one looks, the spirit is upbeat. "The new hotel is designed to meet Harrah's high standard for guest accommodations, and to bring this brand of luxury to a world-class destination like New Orleans."

Top left: Lobby detail.
Top right: Lobby.
Above right: Entrance.

MBH

2470 Mariner Square Loop
Alameda, CA 94501
510.865.8663
510.865.1611 (Fax)
www.mbharch.com

1300 Dove Street
Newport Beach, CA 92660
949.757.3240
949.757.3290 (Fax)

Hilton Grand Vacations Club on the Las Vegas Strip Phase 2
Las Vegas, Nevada

Above: Spa.

Lower left: Grand Staircase.

Lower right: Exterior.

Opposite: Palm-line approach and drive.

Photography: Misha Bruk.

In the city where "what happens here, stays here," the Hilton Grand Vacations Club, located at the north end of the Las Vegas Strip, has given visitors a compelling reason to stay. Conveniently close to gaming, entertainment, and shopping, the 10-acre luxury destination resort includes on-site amenities such as: an oversized 4,000-square-foot spa, both seasonal and heated pools, poolside bars, elegant restaurants, classy lounges, delicatessens and quaint shops. The Club is being developed as a four-phase project, designed in the Desert Art Deco style of the Southwest, that will ultimately include 2,876 keys in four high-rise towers sharing a central lobby and a 14-story parking structure, representing the largest timeshare property in the world. With the completion of Phase II in July 2006, the Grand Vacation Spa opened along with 423 timeshare resort units in a new 38-story tower that connects directly to the central lobby. Already, brisk sales of the new units are paving the way for Phase III, which will bring a 12,000-square-foot restaurant, and amenity deck in addition to the Club's third tower.

MBH

Momo's
San Francisco, California

Located in San Francisco at 760 Second Street across from AT&T Park, home of the San Francisco Giants, Momo's is a 260-seat restaurant, that stands as one of the city's premier dining destinations. Sports enthusiasts arriving for pre- and post-game celebrations have made MoMo's the perfect gathering place. The restaurant's stylish yet casual interiors, including a main dining room, bar and two private rooms, welcome guests with generous leather booths and chairs, rich dark wood finishes, a cherry wood bar, lush reddish-brown carpeting, and soft golden walls accentuated by bright hanging amber light fixtures. Combined with classic American bistro cuisine,

captivating jazz music and congenial service, the warm environment and captivating atmosphere lends Momo's the setting of a world-class saloon. Of course, guests who want to remain outdoors after the game can turn to the restaurant's dining terrace, where bright yellow signature umbrellas open up when the sun gets too hot and heat lamps warm up when the fog rolls in. Momo's is a grand slam for owner Pete Osborne: talented chefs, hearty food and drink, an expansive kitchen, and irresistible dining spaces that attract loyal Giants fans season after season.

Opposite: Main dining room.

Below right: Entrance and dining terrace.

Bottom right: Bar.

Photography: Dennis Anderson.

MBH

Sheraton Princess Kaiulani
Redevelopment Design Competition
Honolulu, Hawaii

The Sheraton Princess Kaiulani first welcomed guests to the former estate of Hawaii's last princess in 1955. How do you update a popular hotel like the Princess, situated a crosswalk away from Honolulu's famous Waikiki Beach, a half century after its completion? To capture the five-acre property's potential, developers at the MacNaughton Group and the Kobayashi Group asked MBH Architects for a redevelopment proposal with one important caveat: Retain the existing hotel tower. MBH's design solution modernizes the 31-story structure, introducing a new high-rise timeshare tower and a new condominium tower, and bringing luxurious amenities such as: a rooftop pool and hot tub, larger guest rooms, floor-to-ceiling glass within each hotel room, new restaurants, spas, retail, and parking for 800 automobiles. The makeover embraces and interprets the Hawaiian lifestyle and culture. Aligned to capture the invigorating Waikiki ocean breeze, the proposal introduces a wave-like swell of alternating balconies and glass curtainwalls, creating a solid-void pattern. The woven patchwork of the thatched roofs and the wave-ripple pattern on the slab system represents the natural ebb and flow of ocean waves. Overall, it's a splendid homecoming for Waikiki's beloved Princess.

Top: Massing model.

Above: Street level perspective.

Opposite: Birdseye view.

Illustration: Courtesy of MBH Architects.

MBH

Taneko Japanese Tavern
Scottsdale, Arizona

The Taneko Japanese Tavern, in Scottsdale, Arizona, reflects a pub-style dining in a Japanese Izakaya also known as a neighborhood tavern. The 5,500-square-foot restaurant is meant to seat 201 guests in an intimate and comfortable setting that includes a 44-seat outdoor dining patio in addition to a 142-seat indoor dining room that encircles a full bar and open kitchen. The design convincingly evokes a rustic setting with finishes such as: leather furniture, hearth stone oven, cherry wood, rich slate and granite, outdoor patio sconces, and custom LED-based fixtures. Illuminating the vibrant Japanese-inspired environment are large, red globe pendant light fixtures that emanate a magical glow that already has the food critics and the public equally enchanted.

Top: Exterior.
Above right: Bar and kitchen.
Right: Dining room booths.
Photography: Larry Falke.

Peter Fillat Architects

Baltimore Office
720 Aliceanna Street
Suite 200
Baltimore, MD 21202
410.576.9310
410.576.8565 Fax

San Diego Office
830 Agate Street
San Diego, CA 92109
858.488.8850
858.488.8855 Fax
http://www.pfarc.com

Peter Fillat Architects

Bear Creek Mountain Resort
Macungie, Pennsylvania

More than a name has changed since skiers took to the rugged slopes of Bear Creek Mountain Resort in Macungie, Pennsylvania in 1968, when the facility opened as Doe Mountain Ski Area. The former day-use facility is now a destination resort, with phase one of its master plan offering 110,000 square feet of new construction on 330 acres, including a ski lodge, 53-room hotel with conference facilities, 183-seat restaurant, 570-seat cafeteria, kitchen, ballroom, indoor pool, daycare center, instruction area, ski rental and retail shop, all designed by Peter Fillat Architects with R.D. Jones & Associates as interior designer. The award-winning project is particularly impressive for its siting, placing the structure on a slope as a series of continuous extensions, like farm buildings in surrounding Berks County, and its use of resources, adding facilities to accommodate more daytime patrons and new overnight guests while conserving energy, water and materials. With a straightforward scheme, locating recreational functions on a long activity axis and hotel functions on a short hospitality axis, and a fresh, contemporary design, featuring Amish-inspired architecture and timeless, rustic interiors, the resort's transformation has been so successful that phase two of its master plan will open in Fall 2007.

Upper Right: Lodge entrance.
Right: Retail shop.
Lower right: Restaurant bar.
Bottom: Exterior showing sloping site.
Photography: Michael Barone, Michael Dersin, Steve Wolf, and courtesy of Bear Creek Mountain Resort.

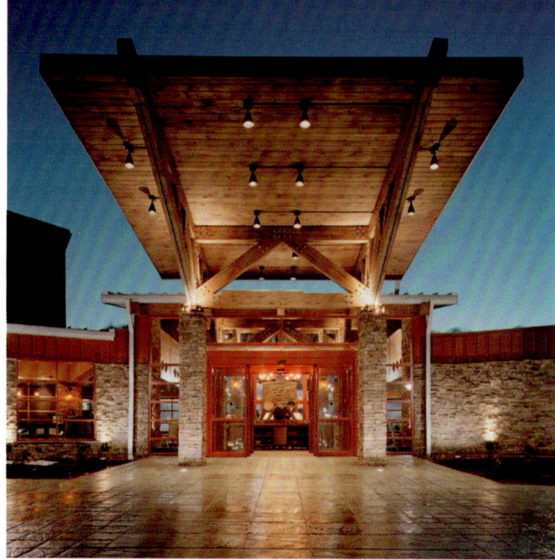

Peter Fillat Architects

St. Louis Renaissance Grand

St. Louis, Missouri

A daunting obstacle faced the building team for a new, 900-room convention headquarters hotel in downtown St. Louis. What should be done with the historic 1904 Statler Hilton, an icon of the women's suffrage movement still occupying the site? Working closely with Peter Fillat Architects, Forrest Perkins, interior designer, Department of the Interior and State Historic Preservation Office, the developer of the St. Louis Renaissance Grand constructed a 23-story addition to the existing structure and a new, adjacent 55,000-square-foot conference center and garage that connects to the hotel via an underground concourse. The result is an award-winning, state-of-the-art facility with national landmark status comprising 918 guestrooms, 10,468 square feet of flexible function space, two ballrooms, 30 meeting rooms, business center, 210-seat restaurant, 98-seat specialty restaurant, coffee bar, lounge bar, health club and indoor pool. Old and new blend seamlessly inside and outside, so the period-style exterior of brick, cast stone, precast concrete, and synthetic stucco is matched by interiors where custom-designed furnishings reflect the structure's historic nature without mimicking a specific period. Indeed, with such attractions as America's Convention Center, Gateway Arch and Busch Stadium just steps away, the hotel offers guests the best of St. Louis, past and present.

Top left: Exterior.
Top right: Façade detail.
Left: Ballroom.
Opposite: Lounge bar.
Photography: Alise O'Brien.

Peter Fillat Architects

Courtyard by Marriott Silver Spring Downtown
Silver Spring, Maryland

Downtown Silver Spring, Maryland, Washington's first suburban shopping district, is a rags-to-riches-to-rags story with a happy ending that has even surprised its own residents. Now that the American Film Institute and the Discovery Channel have made the community their home and a destination for art and entertainment, a wave of development has added new offices, shopping, theaters, restaurants, housing and such hotels as the new, 179-room, 139,200-square-foot Courtyard by Marriott Silver Spring Downtown, designed by Peter Fillat Architects. Courtyard by Marriott is a quintessential urban hotel, skillfully exploiting its site to accommodate guestrooms, meeting rooms, kitchen, restaurant, bar and lounge, public work area, remote lobby, bridge to adjacent public parking, landscaped courtyard and two-level retail space. To maximize street level retailing, for example, the design places a prominent shuttle lobby at street level to take guests to public hotel services on level three. Guests enjoy the use of a landscaped courtyard 40 feet above ground level thanks to a depressed slab and irrigation system. Courtyard's standard interior furnishings have been upgraded, so the lobby displays cherry wood paneling and colorful terrazzo flooring. Travelers seeking an appealing "inside-the-Beltway" hotel will find Courtyard by Marriott holds the keys to Silver Spring.

Top: Exterior at twilight.
Above: Retail storefronts.
Right: Daytime view of exterior.
Opposite: Lobby.
Photography: Michael Dersin.

Peter Fillat Architects

Westin at the Center Convention Headquarters Hotel
Baltimore, Maryland

To give the City of Baltimore a world-class convention headquarters hotel with meaningful public spaces, the developer of the new, 803-key, 628,000-square-foot Westin at the Center Convention Headquarters Hotel and Peter Fillat Architects have created a facility that will simultaneously provide dynamic new venues while reinforcing the city's urban institutions and city life. Westin at the Center focuses on meeting the needs of conventions with 48,000 square feet of meeting space, a 25,000-square-foot ballroom, a 20,000-square-foot café and an outdoor terrace. At the same time, the facility gives guests incomparable lodgings, from the suave, high-tech, 42-story guestroom tower to the welcoming health club and indoor pool, all conveniently located minutes from the Baltimore Convention Center, historic Camden Station and Camden Yards Ballpark, home of the Baltimore Orioles, the central business district, and Inner Harbor's lively mix of culture, recreation and shopping.

Top: Aerial view showing downtown Baltimore and Inner Harbor.

Above: Skyline view.

Right: Exterior with Baltimore Convention Center.

Illustration: Courtesy of Peter Fillat Architects.

sfa design

3888 State Street
Suite 201
Santa Barbara, CA 93105
805.692.1948
805.692.9293 (Fax)
www.sfadesign.com

sfa design

Venetian Resort Hotel Casino
Venezia Tower
Las Vegas, Nevada

Making waves takes a monumental effort at the Venetian Resort Hotel Casino, a 4,027-room tribute to the Queen of the Adriatic that is one of the most opulent mega-resorts in Las Vegas. Yet the new Venezia Tower has achieved nothing less. The 1,100-room, 12-story luxury guest facility, featuring an interior design by sfa design for the lobby, elevator lobbies, guestrooms and suites, concierge lounge, corridors and related public spaces, was deliberately conceived to be a "boutique" hotel-within-a-hotel. It has succeeded so well that it serves as a standalone destination with an identity as unique as the Venetian's famed recreations of such landmarks as the Campanile, Clock Tower and Doges Palace, the Grand Canal Shoppes, a mall with canal and gondolas, the Guggenheim Hermitage Museum and Madame Tussauds Las Vegas. A series of inspired decisions by sfa design and its project team enabled the Venezia Tower to achieve its goal. First, to stage a grand entrance in a lobby located above a parking garage, the front desk faces a lush garden and pool area formally landscaped by Don Brinkerhoff of Lifescapes to resemble a private Roman garden. Guests are invariably surprised and delighted to arrive in this enchanted place above street level. In addition, public areas display a generosity of space matched by a reverence for detail. A dramatic, elliptical rotunda, for example, sets the tone for reception areas with a vaulted ceiling, original murals reminiscent of early 18th-century frescoes, intricate stone patterns, rubbed plaster walls, backlit onyx front desk, and custom-designed light fixtures that reinforce the formality of

architecture and landscape. Nevertheless, the design refrains from excessive ornamentation, eschewing superfluous furnishings much as its historic counterparts did to emphasize symmetry, scale and form. Finally, the 700-square-foot guest suites expand on the already sumptuous standards of the Venetian through Old World luxury and 21st-century convenience. Not only does each room offer gracious, transitional style furnishings that

Above: Corridor and vestibule.

Left: Lobby lounge and bar.

Opposite: Registration lobby rotunda.

Photography: Erhard Pfeiffer.

include wood veneer paneling inset with handcrafted silk damask, wrought iron railings with 22k gold accents, hand-selected marble, meticulously tailored wood and upholstered furniture, and a collection of original art—all beneath a nine-foot ceiling—it also boasts a sparkling, five-fixture bath of marble and gold fixtures, along with three telephones, two 27-inch flat-screen TVs, DVD players, in-room Internet service and every other amenity expected of a luxury hotel. Elegantly gilded as the Venetian is, Fodor's declares the Venezia Tower to be "even posher."

Above: Guest bath.
Left: Guest suite.
Opposite: Main corridor.

sfa design

Fairmont Heritage Place Acapulco Diamante
Acapulco, Mexico

Left: Villa bedroom.

Above: Villa bath.

Below: Villa living/dining area and kitchen.

Opposite: Arrival pavilion.

Photography: Digital Dawgs, Patricia Madrigal Elizondo/Ambientes por Obras.

It's almost as if Hollywood had conjured the site, a sandy white beach skirting the blue waters of the Pacific Ocean in Acapulco, Mexico. Fairmont Hotels & Resorts has developed a new, private residence club where members enjoy worry-free ownership of luxurious private homes in an idyllic setting that seems almost too perfect to be real. Fortunately, Fairmont Heritage Place Acapulco Diamante is very much a reality for families staying in the 50 3,000-square-foot fractional ownership villas and the arrival pavilion, pool pavilion and café that accompany them, all designed by EDSA as architect and sfa design as interior designer. The sleek, contemporary facility reflects both artistic creativity and market research. Focus groups conducted by sfa design revealed that future owners, affluent Mexico City families, wanted a modern, sophisticated and slightly Asian milieu instead of the traditional Mexican-American aesthetic to characterize an environment where indoor and outdoor space intermingled. To satisfy families with children, groups of rooms are casually integrated to create ample opportunities for socializing, interiors are appointed in light, nature-themed, and locally-sourced furnishings mixing Old World Mexican comfort with modern, international style, and subtle lighting mimics the feel of candlelight. In other words, exquisite family vacations flourish here.

sfa design

Fairmont Turnberry Isle Resort & Club
Miami, Florida

Right: Guestroom.
Below: Guest bath.
Bottom: Guestroom.
Photography: Ken Hayden.

Can a fabled grand resort—sheltering such luminaries as Elizabeth Taylor, President Bill Clinton, Billy Crystal and Oprah Winfrey—surpass itself in a renovation meant to reward loyal patrons and attract new clientele? The question is timely and relevant for the 392-room Fairmont Turnberry Isle Resort & Club, in Miami, and sfa design, its interior design firm, as a $100-million renovation of the 300-acre property proceeds. The fast-tracked makeover of guestrooms and suites in three guest towers shows what guests can expect when the entire project is completed. Since each tower is named for a flower, namely Hibiscus, Jasmine and Magnolia, the renovation incorporates each flower into its corridor carpet, assisting guests in navigating the resort and introducing a new visual theme for individual spaces. Overall, the typical guestroom blends existing architectural elements with fresh transitional furnishings, incorporating modern materials, clean lines and neutral-colored fabrics, along with modern amenities, while typical guest bathrooms bring together traditional stone flooring, stone mosaic borders and marble vanity tops with a contemporary, custom-designed vanity, shadowbox mirrors and sleek light fixtures. What do guests think of the facelift? The response is not hard to gauge in the crowded interiors of this living south Florida legend.

Steelman Partners

3330 West Desert Inn Road
Las Vegas, NV 89102
702.873.0221
702.367.3565 (Fax)
www.steelmanpartners.com

Steelman Partners

Harrah's North Kansas City Expansion
North Kansas City, Missouri

Where do fun-seeking Midwesterners play in Kansas City? Besides such attractions as the Kansas Speedway, Country Club Plaza, Truman Sports Complex, Kansas City Zoological Park, and Nelson-Atkins Museum of Art, Harrah's North Kansas City, in North Kansas City, shows up increasingly on visitors' itineraries. The popular hotel and casino, originally built with 181 guestrooms, has gained stature as an attractive and affordable place where plenty of gaming is served along with comfortable guest accommodations, refined restaurants, and a dazzling variety of live entertainment venues. Now, responding to popular demand, Harrah's has added a 211-room, 12-story hotel tower and lobby, along with such amenities as a new casino atrium, casino bar, Toby Keith's I Love This Bar & Grill, Moby's Seafood and Voodoo Lounge Nightclub, all designed by Steelman Partners as architect and interior designer. By linking the style of the new spaces with existing ones, Steelman Partners has brought a convincing sense of continuity to the entire complex and drawn more customers than ever before. In fact, by designing the suave Voodoo Lounge to function as a live performance venue, meeting space, glamorous lounge, and sports bar, the addition has created what is possibly the hottest nightspot in town.

Clockwise from top left: Theater; hotel tower; poker room; tower lobby; Toby Keith's I Love This Bar and Grill.

Opposite: Casino atrium.

Photography: Steve Swalwell.

Harrah's

Steelman Partners

Sands Macao Expansion

Macao, China

China is growing torrentially, and a new, 64,000-square-foot expansion at the Sands Macao, the first Western-operated gaming and entertainment destination in the People's Republic of China, gives this booming Macao property the world's largest casino—covering over 229,000 square feet. The results are impressive enough in quality as well as quantity to captivate day-trippers and high rollers alike. Designed by Steelman Partners, the three-level addition has introduced new gaming areas, lounge/entertainment space and cashiering, motivating guests to explore and use multiple gaming floors without having lost a single day of existing gaming operations. Steelman Partners has shrewdly provided the new gaming areas their own individual design themes to create interest and excitement while maintaining visibility from the main gaming floor. A gold theme on the same level as the existing casino, for example, extends its distinctive look. A bamboo theme on the Fortune (second) level enriches forms, colors and textures. A red theme marks the VIP casino on the Treasure (third) level, and applies traditional Chinese red to wood trim, carpet and furnishings. Commenting on the expansion, Mark Brown, Sands Macao president, has said, "This extension will provide our customers with a diversified range of opportunities to expand their entertainment pleasure."

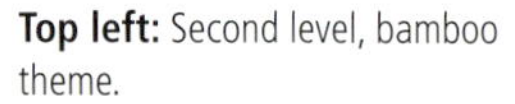

Top left: Second level, bamboo theme.

Left: First level, casino expansion, gold.

Below left: Third level, red.

Bottom left: Second level.

Bottom right: Third level.

Opposite: First level.

Photography: Kam Ieng Lou.

Asian Princes

Steelman Partners

Silverton Hotel and Casino
Seasons Buffet
Las Vegas, Nevada

With an Egyptian-style pyramid, copies of iconic Manhattan skyscrapers, a miniature Eiffel Tower and other exotic structures in Las Vegas shouting for attention, the Silverton Hotel and Casino happily evokes a less frantic, more relaxed and decidedly rustic world in its turn-of-the-century image of an Adirondack-style mountain lodge. Guests who flock to its 300 rooms, numerous restaurants and lounges, 60,000-square-foot casino, salt-water aquarium and other attractions, appreciate its distinctly down-home flavor. Showing how well Silverton knows its customers, Seasons, a new, 280-seat, 11,000-square-foot buffet service, designed by Steelman Partners, blends old-time charm with modern detailing to give guests a memorable dining experience quite unlike what other local buffets offer. Its spacious dining room, framed in dark timber columns, beams, arches and brackets, illuminated by wrought-iron-and-glass chandeliers, ceiling coves and other lighting sources, and graced by earth-toned walls, carpet and upholstery, feels like a friendly family restaurant, an effect heightened by the use of low visual barriers to create intimate smaller areas. There are even six live-action serving stations providing a dash of showmanship. For guests who like the practicality of a buffet served with restaurant ambience and entertaining live action, Seasons is always in season.

Right: Dining room.
Bottom: Entrance.
Photography: Kory Pittman.

Steelman Partners

MGM Grand
MGM Race and Sports Book, Poker Room, and Centrifuge Bar
Las Vegas, Nevada

Superlatives have come easily to the 5,044-room MGM Grand hotel and casino in Las Vegas since its 1993 opening, a point reiterated whenever guests approach its signature 100,000-pound bronze lion statue. Reluctant to rest on its laurels, however, MGM Grand continually develops new attractions such as the trio of new spaces, designed by Steelman Partners, that have transformed the seldom-visited domed area at the main entrance off Las Vegas Boulevard. The 5,300-square-foot MGM Race and Sports Book places guests in a state-of-the-art arena, featuring a 150-seat Sports Book that comes with 36 60-inch plasma televisions, eight 42-inch plasmas, six electronic display boards and 17 separate race and sports betting counters. Also, there is a 49-seat Race Book where each guest has an individual 15-inch color television to personally select events for monitoring. The 5,200-square-foot Poker Room capitalizes on the game's growing popularity with 22 tables in a dramatically lighted room boasting the advanced accommodations enthusiasts expect. The 80-seat, 3.600-square-foot Centrifuge Bar beckons guests to a open gallery of gleaming concentric rings that glitters like a space ship with its bar at the center of the circular room. Guests can easily find these new attractions by following in the crowds.

Top: MGM Race and Sports Book.

Upper left: Centrifuge Bar.

Left: Poker Room.

Photography: Jeffrey Green Photography.

Steelman Partners

Macao Studio City
Macao, China

Real estate developers are racing to house China's red-hot economy. Macao Studio City, a joint venture of ESun Holdings and New Cotai LLC, will do nothing less than establish Asia's first leisure resort to integrate television and film production facilities with retail shopping, gaming and world-class hotels on a 35-acre site in Macao by 2009. Providing master planning, architecture, interior design, graphics and lighting design for the 2,010-room, 3.66-million-square-foot project, Steelman Partners has created a timeless, modern design that resolves an unusual request from the developers: create four hotels with separate lobbies that are vertically integrated in one guest tower yet could be marketed independently. Each major component is not unlike a small city. The entertainment portion, for example, will include a 2,300-seat theater, 4,700-seat arena, standalone television and film production facilities, and 1.4-million-square-foot Studio Retail™ shopping center. The four hotels, to be sheathed in a glass tower with programmable LED lighting that each hotel can custom tailor to express its identity, will incorporate such separate and shared spaces as reception areas, lobbies, lounges, guestrooms, restaurants, kitchen, ballroom, bars, theater, pool, and gaming facilities. In fact, Macao Studio City should easily live up to its name.

Top right: Detail at street level.

Above: Overall view.

Illustration: Courtesy of Steelman Partners.

Stephen B. Jacobs Group

Andi Pepper Interior Design

677 Fifth Avenue
New York, NY 10022
212.421.3712
212.752.4819 (Fax)
www.contactus@sbjgroup.com

Stephen B. Jacobs Group

Andi Pepper Interior Design

Hotel Giraffe

New York, New York

Small gestures handled with exceptional skill and style get noticed even in big cities, and the new, 12-story, 73-room Hotel Giraffe, designed by Stephen B. Jacobs Group and Andi Pepper Interior Design, is enchanting guests in midtown Manhattan despite or because of its size. The award-winning luxury boutique hotel, located in the charming Madison Park neighborhood, is a new structure designed to evoke the understated French Art Moderne style of the 1930s. Like its namesake, the hotel relies on grace, strength and beauty for impact, blending with neighboring buildings yet drawing attention with an urbane façade of brick and cast stone that features Juliet balconies with French doors for all guest rooms and a summit topped with a barrel vaulted dormer, open arcade and rooftop garden. Similar attention to detail

continues inside, where the lobby, 52 guestrooms, 21 suites, penthouse hospitality suite, restaurant and bar honor the vision of legendary French designer Jacques-Emil Ruhlman. Shrewdly, the design team and owner Henry Kallen have established an intimate oasis that cossets guests in quiet luxury, spacious accommodations, high ceilings, fine furnishings, and two stately public spaces, the lobby and penthouse suite, where baby grand pianos await their sophisticated ladies and gentlemen.

Below left: Lobby.

Below: Exterior.

Below right: Rooftop garden.

Bottom left: Penthouse suite.

Below middle right: Guest bedroom.

Bottom right: Suite living room.

Photography: Bruce Buck.

Stephen B. Jacobs Group
Andi Pepper Interior Design

Library Hotel
New York, New York

Above: Guest bath.

Right: Exterior.

Opposite top: Lobby.

Opposite bottom left: Bar.

Opposite bottom middle: Second floor lounge.

Opposite bottom right: Guestroom.

Photography: Bruce Buck.

Patience and Fortitude, the beloved stone lions guarding the New York Public Library at Fifth Avenue and 42nd Street, have surely noticed the newcomer, the Library Hotel, at the nearby corner of Madison Avenue and 41st Street. Created through a creative adaptation of a 12-story, landmark 1912 brick and terra cotta structure that is just 25 feet wide, the award-winning, 60-room facility, designed by Stephen B. Jacobs Group and Andi Pepper Interior Design, is a luxury hotel with the ambiance of a private club and a literary theme invoking the Dewey Decimal System. Intensive effort went into converting the building for hospitality, beginning with the cleaning and restoration of the ornate façade and its nine-story, five-foot-deep copper bay window, and a gut interior renovation including an MEP upgrade and reorganization of floorplates. To accommodate a lobby, restaurant, second floor lounge and sitting room, bar and business center, guestrooms, and rooftop fireplace lounge, sitting rooms and terrace, the design shifts the entrance from Madison Avenue to the wider frontage on 41st Street, incorporates extensive built-ins, and features interiors where elegant, comfortable furnishings contrast dark woods against light fabrics and leathers, and—honoring the erudite neighbor—some 6,000 books.

Stephen B. Jacobs Group
Andi Pepper Interior Design

Hotel Gansevoort
New York, New York

Below left: Exterior.
Below: Front desk.
Below right: Reception/lobby.
Bottom left: Entrance.
Photography: David Joseph, OTHERS TO BE IDENTIFIED.

Just a decade ago New Yorkers might have ridiculed anyone asking directions to a sophisticated, 187-room, 14-story contemporary luxury hotel in the historic Meatpacking District. Even today, the neighborhood

remains a meat wholesale market, although a young, stylish and entertainment-oriented crowd is making steady inroads through trendy restaurants and chic boutique shops. But the Hotel Gansevoort stands as a compelling symbol of one of Manhattan's hottest emerging neighborhoods. Owing to the determination of the Gansevoort's developer to proceed with financing and construction in the wake of 9/11, the hotel opened when the city was eager to resume normal life. The timing was fortuitous in more ways than one, permitting the lofty, as-of-right structure, designed by Stephen B. Jacobs Group and Andi Pepper Interior Design, to be completed before a subsequent zoning change imposed severe height restrictions on new construction. As the enduring popularity of the hotel demonstrates, the hotel's understated yet stylish contemporary environment —a double-height lobby, restaurant, conference area, meeting rooms, guestrooms, and rooftop pool and bar within a cool, crisply detailed, zinc-colored metal façade—is exactly what its youthful clientele wants. Distinctive

Stephen B. Jacobs Group
Andi Pepper Interior Design

Right: Rooftop pool.
Below: Guestroom.
Bottom left: Guest bath.
Bottom right: Guest desk.

detailing of the architecture and interior design helps make the Gansevoort a magnet for social activity, whether guests are arriving for lodging, dinner or cocktails. For example, the 45-foot-long rooftop pool is not the only one of its kind in Manhattan, but teaming it with a sleek bar/lounge and elaborately landscaped garden gives it a special allure, particularly on a summer evening. Sophisticated furnishings, such as a brilliantly colored rug in the lobby custom designed by Andi Pepper, comfortable minimalist furniture, such timeless finishes as textiles, leather, marble, slate and stainless steel, and internally illuminated, 18-foot glass-sheathed columns that change their colors throughout the day and night, give the interiors their own unique character. Bay windows and intimate sitting areas in all guestrooms create a variety of appealing spatial experiences that are quite unlike what typical hotels offer. Perhaps the only aspect of the Gansevoort that rivals its décor, appropriately enough, is its service, described by *Condé Nast Traveler* "remarkably good for a style-conscious boutique hotel."

Studio GAIA Inc.

601 West 26th Street
Suite 415
New York, NY 10001
212.680.3500
212.680.3535 (Fax)
www.studiogaia.com

Studio GAIA Inc.

W Mexico City Hotel
Mexico City, Mexico

Far left: Entrance.

Left: Stairs to Red Lounge.

Below: Motor Lobby and The Whiskey.

Opposite: Red Lounge.

Opposite bottom left: Steps from Motor Lobby to The Whiskey

Opposite bottom right: The Whiskey.

Photography: Jaime Navarro.

The new W Mexico City was obviously meant for Mexico City's trendy, upscale Polanco neighborhood. After all, the global jetsetters who patronize Polanco's chic boutiques, gourmet restaurants and such institutions as Chapultepec Park and the National Auditorium are drawn from the same clientele that registers in other W Hotels from Starwood Hotels and Resorts Worldwide. Indeed, the 250-bed, 182,920-square-foot W Mexico City, the first W Hotel in Latin America, has instantly become a popular addition to the neighborhood's business and social life, animated by guests attracted to the innovative and exciting interior design created by Studio GAIA. One glimpse at such facilities as the lobby lounge, three restaurants, including The Living Room, Solea and The Terrace, aWay Spa and 250 guestrooms and suites dramatically demonstrates that the hotel has encouraged its design team to be as imaginative and daring as possible. As a result, boundaries between public and private, interior and exterior, and hotel and residence are provocatively

Studio GAIA Inc.

blurred, the lobby lounge boldly reveals the hotel's multi-level public areas—including the red lounge that leads directly to the second-floor restaurant and dazzling glass box enclosing its private dining room—and ultra-modern guestrooms indulge guests with all-white signature beds, cherry red walls and walk-in shower rooms. High fashion does not prevent the W Mexico City from being friendly and amenable to guests and visitors, however. The striking contemporary design that creates theatrical spectacles also provides residential-style convenience and comfort in public and private accommodations through a combination of intense, warm Mexican colors, timeless local materials such as lava stone, terrazzo, and Cantera Paloma stone, stylish modern furnishings, and sophisticated architectural lighting. Guests seeking an appropriate dining venue, for example, can choose the urban flair of The Living Room's cozy couches in the

Top: Solea communal table.

Above far left: Solea banquette seating.

Above left: Cocoa Bar.

Far left: Temazcal area in a Way Spa.

Left: Relayation pool in a Way Spa.

Opposite: Solea Laja Private Room.

Studio GAIA Inc.

lobby lounge, the dynamic Solea's multi-level structure, or spectacular evening views from The Terrace's lofty perch. When it's time for privacy and repose, they have the choice of indulging in luxury guestrooms, palatial Loft rooms or Presidential/ Extreme WOW suites, all equipped with such amenities as showers with massage jets or spacious tubs, hammocks in all shower rooms, oversized desks and IT services, state-of-the-art entertainment

Top left: Corridor and room entry.

Top right: Loft.

Above right: Loft bath.

Above: Guest bath.

Right: Guest bedroom.

Studio GAIA Inc.

systems, and the feather and down comforts of the W Designer Bed. Starwood has responded to the success of the W Mexico City as a vision of sophisticated, modern Mexico by describing its interior design firm as "true visionaries who absorb and interpret culture and design rather than just echo the latest trends."

Top right: Presidential Suite master bedroom.

Right: Presidential Suite bath.

Far right: Presidential Suite bedroom.

Bottom right: Presidential Suite living room.

tonychi and associates

20 West 36th Street
9th Floor
New York, NY 10018
212.868.8686
212.465.1098 (Fax)
www.tonychi.com

tonychi and associates

Namu/LIQUID Bar and Kitchen
W Seoul-Walkerhill
Seoul, South Korea

Left: Sake bar at Namu/LIQUID Bar.

Bottom left: Main dining room at Kitchen.

Bottom middle: Dining room at Namu/LIQUID Bar.

Bottom right: Private room at Kitchen.

Opposite: Entry "maze of garden hedges" at Kitchen.

Photography: Jason Lang, Suk-Joon Jang.

It's paradise for trendy and sophisticated diners in the capital of the dynamic nation that turned Hyundai, Samsung and LG into global brands. Impressive new restaurants are opening throughout Seoul, with two tempting possibilities being Namu and Kitchen, stylish modern establishments designed by tonychi and associates at the recently completed W Seoul-Walkerhill, a 253-room hotel 15 minutes from downtown. Each restaurant reflects its cuisine and service as well as the chic, ultra-modern hotel. Updated Asian fare is served at the 126-seat, 6,780-square-foot Namu in a design interpretation of a Japanese "hillside barn," and includes Western-style dining rooms, private tatami room, and sake bar. A transparent layering of imagery encircles the lofty, heavy-timbered environment with woods and hills. Since wood-fired, home-style and organic dishes are the specialties of Kitchen, the 338-seat, 6,000-square-foot restaurant is designed as an urban vision of a "chic country house." To welcome guests, the design leads them through a symbolic "maze of garden hedges" to arrive at a main dining room where a long, stone-topped communal table is flanked by banquette seating and private niches as well as a Western-style private room. Whichever restaurant guests choose, the décor is as delectable as the cuisine.

tonychi and associates

Wolfgang Puck MGM Grand
MGM Grand
Las Vegas, Nevada

Right: Exhibition kitchen counter and seating.

Below: Main dining room.

Bottom right: Seating before curtain/space divider.

Bottom far right Decorative backlighted wall panels.

Opposite left: Wine cellar/ private dining room.

Photography: Eric Laignel.

Guests love the legendary 5,044-room MGM Grand in Las Vegas for such dynamic attractions as its 170,000-square-foot casino, but they also happily surrender to its calm, soothing amenities at the right moments. That's why a restaurant resembling a traditional California bungalow, complete with patio, hedges and plaid awning, flourishes inside the new, 200-seat, **0,000**-square-foot Wolfgang Puck MGM Grand, designed by tonychi and associates. In a relaxed milieu celebrating the California patio lifestyle, guests can enjoy such Wolfgang Puck entrées as foccacia with veal meatballs and locatelli or truffled potato chips with blue cheese. "The design combines the energetic feel of the beach lifestyle and the cool beauty of a California garden," explains Tony Chi, principal of tonychi and associates. Guests seated in the expansive main dining room, which sustains a refreshingly convincing outdoor ambiance, or smaller dining areas, including a private dining room/wine cellar, can easily imagine that the interiors of wood, stone and drywall, furnished with clean, contemporary residential-style furnishings and illuminated with soft direct and indirect lighting and backlighted decorative panels, have momentarily transported them to the Golden State. Isn't that what good design is supposed to do in Las Vegas?

tonychi and associates

InterContinental Genève
Geneva, Switzerland

While most international hotels in Geneva, Switzerland line the shores of beautiful Lake Geneva, the lakefront's allure did not deter the five-star, 328-room InterContinental Genève from choosing a radically different site in 1964. Interestingly, being one mile from the lake yet steps from the European headquarters of the United Nations has not deterred such VIPs as Sophia Loren, Kofi Annan, Roger Federer and hundreds of heads of state from checking into the InterContinental Genève. To burnish the hotel's cachet, a comprehensive renovation is currently underway, designed by tonychi and associates, dramatically updating public areas and now visibly transforming guestrooms. "The objective was to provide the hotel with a completely new look, modern but classical, and at one with the natural beauty and sophisticated international presence of the city," observes Tony Chi, principal of tonychi and associates. The ambitious makeover embraces all key areas, from the entrance and reception to such public facilities as the salon, 16-room conference center, gymnasium, ballroom, Woods restaurant, two bars, fitness center and Clarins spa—as well as 226 guestrooms and 102 suites. To update the hotel while reaffirming its ties to Swiss life and global affairs, the design creates a dynamic synthesis between historic styles and classic modernism that gives individual areas their distinctive character. At the entrance, for example, a remodeled façade with new canopy, limestone columns, bronze doors and giant bamboo plants magnifies the visual impact beyond actual dimensions. The reception area has shed its past as a shopping arcade to become the stately Great Hall, graced by monumental columns, grand staircases and a two-story limestone fireplace at the Bar des Nations. In Woods, fine wood cabinetry and a huge wooden table

Right: Lobby/reception desk.
Below left: Great Hall.
Below right: Exterior entrance.
Opposite bottom right: Custom crafted table at Woods restaurant.
Photography: Eric Laignel.

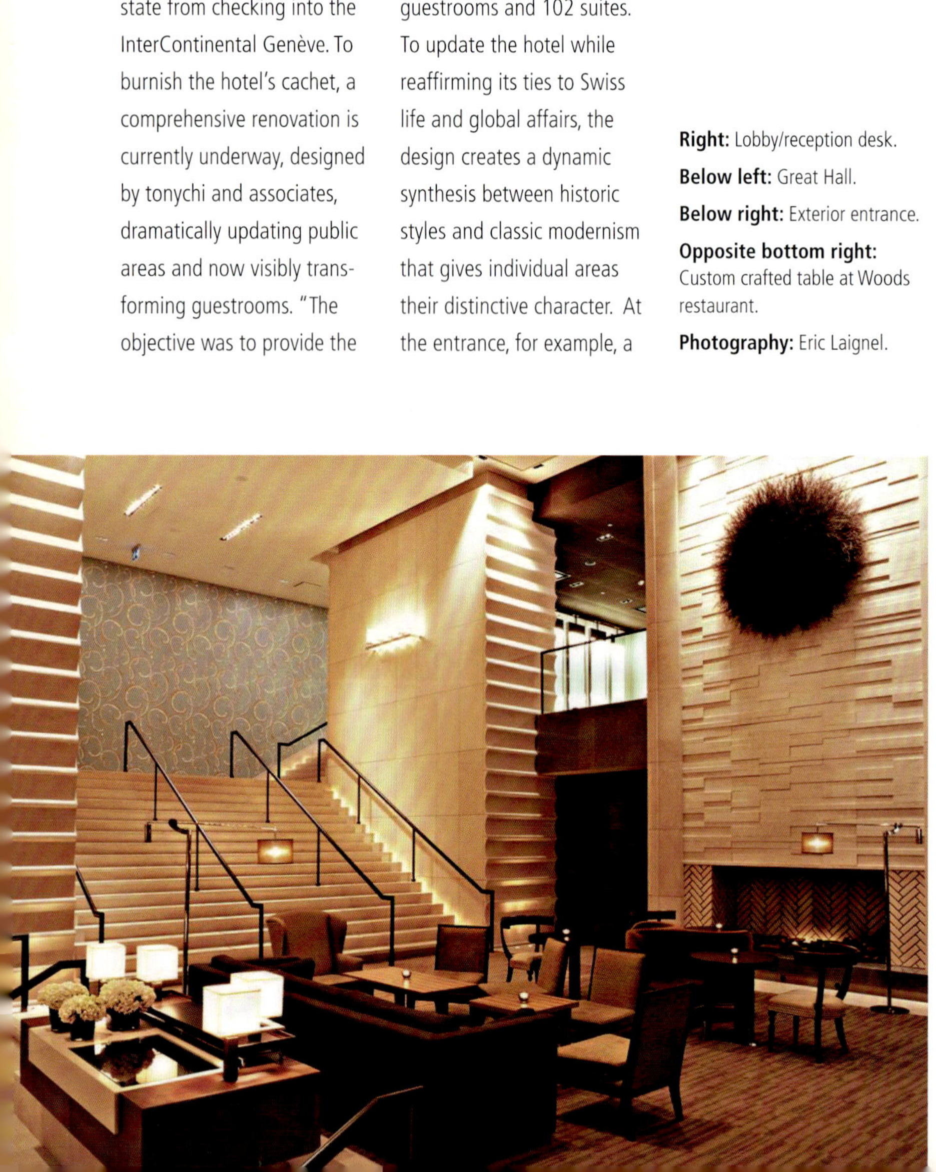

made of tree trunks at the entrance create a large, airy and informal setting for fine seasonal cuisine. Soon, new interiors will showcase guest accommodations featuring contemporary design with sleek, stylish furnishings, a cool, neutral color palette, and subtle lighting, along with views of Lake Geneva, the Alps and/or the Jura Mountains. When the renovation is complete, this unofficial destination for visiting diplomats and celebrities will surely continue its remarkable, four-decade run.

Clockwise from top left to right: Ballroom; Woods Restaurant main dining room; Bar des Nations; conference room; salon.

WATG

IRVINE OFFICE
8001 Irvine Center Drive
Suite 500
Irvine, CA 92618
949.574.8500
949.574.8550 (Fax)

LONDON OFFICE
Alexandra House
6 Little Portland Street
London W1W 7JE
011.44.207.906.6600
011.44.207.906.6660 (Fax)

ORLANDO OFFICE
315 E. Robinson Street
Suite 350
Orlando, FL 32835
407.298.9484
407.298.9184 (Fax)

SEATTLE OFFICE
1201 Western Avenue
Suite 350
Seattle, WA 98101
206.275.2822
206.275.0692 (Fax)

HONOLULU OFFICE
700 Bishop Street
Suite 1800
Honolulu, HI 96813
808.521.8888
808.521.3888 (Fax)

SINGAPORE OFFICE
112 Robinson Road
Unit #07-03
Singapore 068902
65.6227.2618
65.6227.0650 (Fax)

www.watg.com

WATG

Grecotel Cape Sounio
Attica, Greece

The ancient Greeks who built the venerable Temple of Poseidon in 440 B.C. at Cape Sounio in Greece's Attica region could not have foreseen that their creation would become a backdrop for the newly renovated Grecotel Cape Sounio. Nevertheless, the temple's Doric columns are as visible from guests' windows as the Aegean Sea. Spectacular scenery is just one way Grecotel Cape Sounio distinguishes itself as a luxury resort. Its 154 bungalows and such extensive public facilities as a lounge and bar, ballroom, conference center, restaurant, spa, indoor and outdoor pools, and tennis courts now grace a natural amphitheater. The project is part of a 20-acre estate in a pine forest adjacent to Sounio National Park, signaling the rebirth of a resort originally completed in 1973 with new architecture designed by WATG. Inspired by the magnificent site, the design team has created an award-winning contemporary complex that features terracotta-clad Neoclassical buildings, including bungalows of varying size and complexity with private gardens and pools, landscaped terraces adorned with intricate water features, contemporary and traditional furnishings equipped with state-of-the-art amenities, and a glass boardwalk leading to vistas of the site's own archaeological ruins—a resort fit for Poseidon.

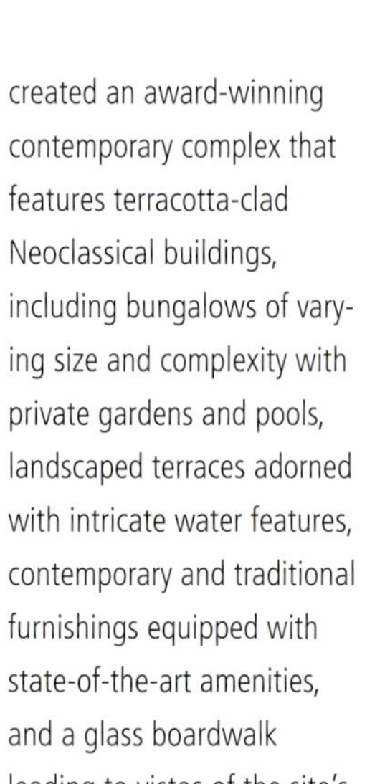

Right, top to bottom: Overall view of main facilities; Bungalows with private gardens and pools; Private pool; Water feature seen against Aegean Sea.

Left: Lounge bar.

Opposite: View from hotel of Temple of Poseidon and Aegean Sea.

Photography: Heinz Troll, courtesy of Grecotel.

Shangri-La's Barr Al Jissah Resort & Spa
Muscat, Oman

Left: Exterior with pool and "lazy river."

Bottom: Restaurant.

Opposite: Courtyard.

Photography: George Mitchell, Group Photographer.

For all the attention its flashy Gulf States neighbors Dubai, Bahrain and Qatar are attracting in their drive towards modernity, the Sultanate of Oman is also updating its physical environment. Oman is proceeding in its own distinctive manner, however, balancing the excitement of dazzling new developments to foster tourism with respect for the tradition of its panoramic deserts, rugged hills and dramatic coastlines, well-preserved fortresses and other historic sites, and tolerant, friendly population of 2.3 million citizens. A shining example of how Oman is embracing the 21st century is the new, three-hotel, 680-key luxury resort and spa, Shangri-La's Barr Al Jissah Resort & Spa, in the capital city of Muscat, designed by WATG. The challenge for the design team was to plan the nation's first integrated destination resort to cater to both business and leisure guests without harming a pristine, 124-acre site endowed with historical and environmental significance. Considering the ambitious building program

for Shangri-La's third Middle Eastern hotel, the design's ability to fit everything into the side of a mountain and along a sandy bay (the Barr Al Jissah) and adjacent headland is quite remarkable. Among the Shangri-La's extensive features are the five-star, 302-key family-oriented Al Waha ("The Oasis") hotel, five-star, 198-key business-oriented Al Bandar ("The Town") hotel, and six-star, 180-key sophisticated Al Husn ("The Castle") hotel, along with a CHI Spa Village, conference center with ballroom and 11 meeting rooms, 1,000-seat open-air amphitheater, 19 food and beverage outlets (including six main restaurants, seven casual dining outlets and pool bars, three lobby lounges, two bars and a nightclub), three swimming pools (two joined by a specially designed "lazy river"), four tennis courts, Omani Heritage Village Museum, landscaped gardens and private beach. Owing to the careful organization, coordination and integration of its complex elements, along with an undeniable aesthetic appeal, Shangri-La has been an unqualified success from its opening. Even the resort's sensitive archaeological sites, bearing treasures 2,000 years old, and breeding coves for rare sea turtles have been seamlessly folded into the guest experience as sources of pleasure and enlightenment for guests and hosts alike.

Top left: Pre-function room.

Top right: Portal overlooking Barr al Jissah.

Bottom: Evening view of exterior.

Opposite: Pavilion.

Photography: Eric Laignel.

WATG

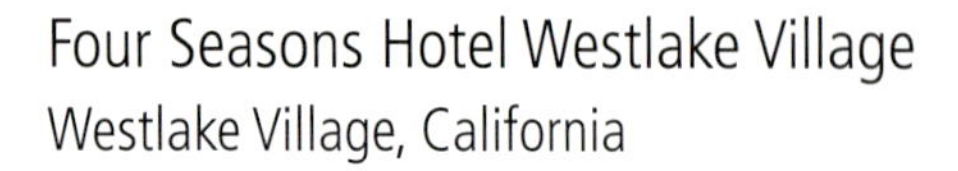

Four Seasons Hotel Westlake Village
Westlake Village, California

Right: Spa exterior.

Far right: Lobby lounge.

Bottom right: Hampton's Restaurant.

Opposite bottom left: Indoor pool.

Opposite bottom right: Guestroom.

Photography: Barbara Kraft, Peter Vitale.

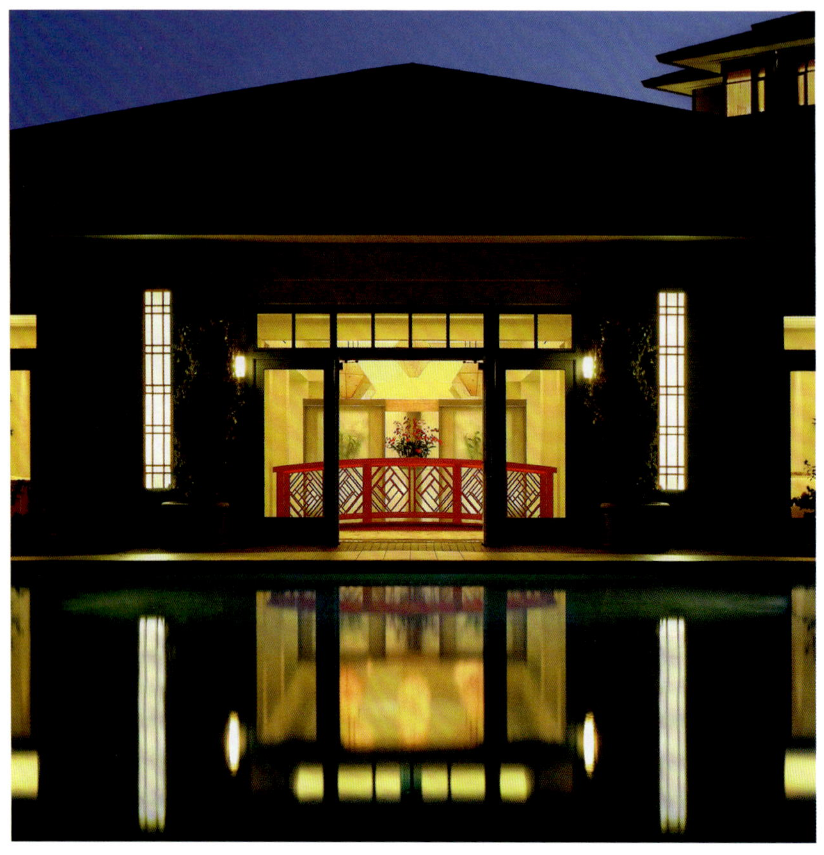

While active, accomplished and affluent people seek to attain healthier and more balanced lives, they often search in vain for world-class facilities addressing every aspect of health and wellbeing under one roof. Happily, residents of greater Los Angeles now have the new, 270-room, 476,317-square-foot Four Seasons Hotel Westlake Village, in Westlake Village, California, designed by WATG in collaboration with Arnold C. Savrann, AIA, design architect. Not only does it teach guests how to better their lives, it nurtures them within a five-star, six-story hotel offering a health and wellness center, spa, medical clinic, conference center, five restaurants, two lounges, Kid's Club, and television studio, set on nine acres of gardens and water features adjacent to the California WellBeing Institute and California Health & Longevity Institute. What enables Four Seasons Hotel Westlake Village to function as one well-integrated environment is a carefully structured scheme with traditional floor plans and classic Georgian architectural styling that achieves a cozy residential ambiance using warm colors, rich materials, period-style furnishings, custom artwork, and sophisticated lighting. Happily, guests will also find such delightful surprises as an original Chinese pagoda and expansive waterfall. Shouldn't living well bring pleasures as well as responsibilities?

WATG

Atlantis, Paradise Island Resort
Phase III, The Cove Atlantis and The Residences at Atlantis
Nassau, The Bahamas

Left: Exterior of Phase III.

Below: Marina.

Bottom: Living room of suite

Opposite: Lobby in The Cove Atlantis.

Photography: Courtesy of Kerzner International Resorts.

Atlantis, Paradise Island Resort, the vast, colorful and endlessly entertaining resort and casino on Paradise Island in Nassau, The Bahamas, has done it again, giving travelers another tempting reason to come ashore: Phase III, The Cove Atlantis and The Residences at Atlantis, designed by WATG with HKS as architect of record. The goal of the expansion, comprising The Cove Atlantis, a 600-suite, 820,000-square-foot hotel tower and low-rise structure, The Residences at Atlantis, a 500-room, 610,000-square-foot condominium residential hotel, and a 30,000-square-foot spa, typifies the ambitions of Atlantis, the crown jewel of Kerzner International Resorts. In effect, Phase III is meant to epitomize the upscale Bahamas resort lifestyle. Its fresh design employs a refined, modern architectural vocabulary to balance the theatricality of Phase II, starting with a contemporary exterior featuring "spire" elements that evoke the Atlantis symbol. Inside, strong, simple architectural interiors are expressed in a refined palette of tropical colors, classic materials such as stone, plaster, glass and hardwoods, sophisticated yet lighthearted "tropical glam" furnishings, and an array of lighting. Attention to detail is unflagging despite Phase III's scale, from the moment guests arrive in the lobby and behold the stunning, vaulted nave soaring above.

WATG

Forty-Seven Park Street Marriott Grand Residence
London, United Kingdom

"When a man is tired of London," noted Dr. Samuel Johnson, the great English 18th-century man of letters, "he is tired of life." Fortunately, business and leisure travelers who share his sentiment now have a luxurious alternative to hotels, condominiums and seasonal rentals: Forty-Seven Park Street Marriott Grand Residence. This stately Edwardian-style town house, built in 1926 as a private residence for the 1st Baron Milford in London's exclusive Mayfair neighborhood, has been converted from a hotel into a unique home of 49

one- and two-bedroom guest suites for full or fractional ownership. To create a five-star residential-style hotel including a reception lobby, gymnasium and health spa in addition to the guest suites, the interior design by WATG combines the building's finest historic details with Neoclassical and Regency motifs to produce distinctive residences and common areas. A color palette of beiges, creams, reds and blues, such timeless materials as mahogany, brass, marble, granite and plaster, period-style furnishings, and deluxe kitchen and bath appliances and fixtures give the Marriott Grand Residence a luster that would have pleased such discerning former Mayfair luminaries as Winston Churchill, Henry James and various members of the Royal Family.

Opposite: Suite living room.
Below middle: Exterior.
Below right: Reception lobby.
Bottom left: Guest suite bath.
Lower right: Staircase.
Bottom right: Suite bedroom.
Photography: Courtesy of Forty-Seven Park Street Marriott Grand Residence.

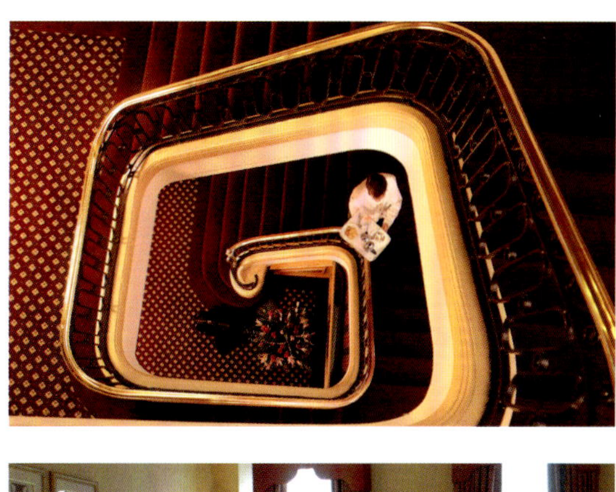

WATG

Hilton Sanya Resort & Spa
Sanya, China

Yalong Bay, a crescent shaped curve of white sand on the southernmost tip of Hainan Island off the southern coast of China near the city of Sanya, is such a beautiful tropical setting for vacationers that the challenge confronting the building team for the new, 501-room, 645,000-square-foot Hilton Sanya Resort & Spa was how to develop a great resort without spoiling it. Working in a close collaboration, WATG, architect, Chhada Siembieda Leung Ltd., interior designer, Belt Collins International, landscape architect, and China Jin Mao Group Co., Ltd., developer, have produced a design that reduces the project's mass and detail, respects the natural environment, and celebrates the architecture

of southern China. Consequently, such facilities as the 32,000-square-foot conference center, six food and beverage outlets, kids club, entertainment center, spa and fitness center, guestrooms and bungalows have flourished within and around the grand yet understated structure of stone, stucco, wood and steel under a clay tile roof. In fact, the award-winning interpretation of traditional regional architecture has proved especially auspicious. From the vast open lobby, dominated by Chinese-style pillars and exposed roof, to the sleek communal spaces and stylish guestrooms, guests are obviously enthralled by everything around them.

Below: Exterior with reflecting pool.

Opposite top left: View of overall site.

Opposite bottom left: Lounge.

Photography: Courtesy of Hilton Sanya Resort & Spa.

WATG

The Lake Resort
Vilamoura, Portugal

It's not every day that you can stumble upon a magnificent, five-star, 196-room resort hotel and 95-unit luxury residential development hidden by an ecological reserve between a marina and the sea. However, that's what guests and buyers are doing by finding their way to The Lake Resort, nestled within the center of Vilamoura, a leading golf and leisure destination in the Algarve region on Portugal's southern coast. For WATG, the architect of The Lake Resort, the key to the project's award-winning design was to integrate the architecture with the natural environment, highlighted by an artificial lake before the building that connects directly with the sea, promoting the renewal of the marina's water and maintaining the reserve. Equally important for guests, of course, is the beauty, versatility and comfort of the hotel and residences, and the architecture is no less conscientious about providing a superior environment for such facilities as the guestrooms, three restaurants, three bars, spa, fitness center, indoor and outdoor pools, mini-club, library, and boutique shops. The design combines traditional Portuguese architecture with classical Roman themes to make guests feel they're visiting a grand Portuguese finca or country mansion—within an ecological reserve, no less.

Right: Lounge.
Below left: Exterior.
Below right: Rotunda.
Photography: Amorim Turismo.

WESTAR Architects

701 Bridger Avenue
Suite 400
Las Vegas, Nevada 89101
702.878.0000
702.878.8430 (Fax)
www.wagnarchitects.com

WESTAR Architects

Bally's Atlantic City Players Club
Atlantic City, New Jersey

Seasoned players enjoy taking intermissions from intensive sessions of gaming for dining, entertainment, and other forms of relaxation. Where to go? The new, single-floor, 19,000-square-foot Players Club at Bally's Atlantic City, in Atlantic City, New Jersey, designed by WESTAR Architects, extends a warm and friendly welcome to some 425 guests en route to or from the casino. To unwind, guests may proceed from the reception area to the 40-seat bar, Americana Lounge, and various other lounges and dining rooms. What makes the facility so appealing is its success at breaking a large floor area into individual rooms that promise unique spatial experiences and luxurious appointments. Such materials and furnishings as finely detailed ceilings of copper, gold and copper leaf, and wood beams, richly embossed wallcoverings, framed mirrors, classic wood and leather upholstered furniture in traditional American styles, four working fireplaces and a wide assortment of artifacts are combined to create environments with a convincing sense of place. Adding to the total effect is a color palette of burgundy, gold, beige, copper and purple, as well as a lighting scheme employing over 30 different light fixtures. Is anyone surprised that guests consider the Players Club to be Bally's most beautiful club?

Top left: Bar.

Top right: Bar and dome.

Right: Seating area.

Opposite: Fireplace lounge.

Photography: Darius Kuzmickas.

WESTAR Architects

Caesar's Atlantic City
Diamond Lounge
Atlantic City, New Jersey

Right: Seating group with Sconces.

Below: Fireplace lounge.

Opposite top left: Coffered sitting area.

Opposite top right: Buffet area.

Opposite Bottom: Reception.

Photography: Darius Kuzmickas.

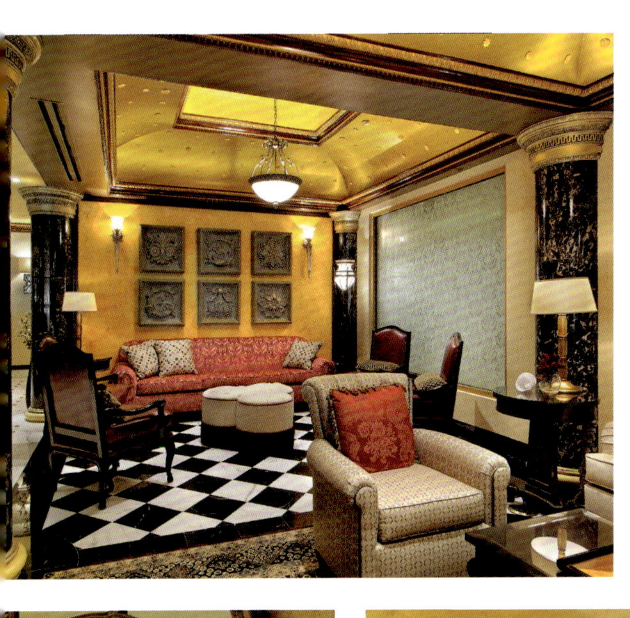

The allure of an opulent, traditional European-style villa awaits high rollers at Caesar's Atlantic City, in Atlantic City, New Jersey, thanks to the opening of the new, single-floor, 14,000-square-foot Diamond Lounge. Designed by WESTAR Architects to offer 345 guests the equivalent of their own, personal space, the Diamond Lounge transforms an odd shaped floor area into rooms of varying sizes and proportions, much as a private residence would. Indeed, the reception area, fireplace lounge, bar, buffet, and multiple lounges are outfitted with exquisite materials and furnishings such as marble, gold leaf, book-matched onyx, crotch-cut mahogany, tile mosaics and furniture in historic styles, embracing a color palette of gold, blue, red, beige and black, to give guests the impression they are visiting the home of a wealthy and sophisticated family. Thus, along with the usual grand salons that host dozens of guests, there are intimate seating areas suitable for three or four couples at the most. Each room or area offers its own interpretation of the resplendent, Old World theme, and uses genuine materials—no faux marble or gold is present—to render its effects, reinforcing the Caesar's brand by pampering high rollers as only Caesar's Atlantic City can.

WESTAR Architects

Trump Plaza Hotel & Casino
Atlantic City, New Jersey

Above: 24 Central's Restaurant entrance.

Right: Casino entrance.

Bottom right: Fortune's Restaurant.

Opposite top: Crystal River.

A much-anticipated remodeling program to bring new glamour, elegance and sparkle to the 39-story, 762-room Trump Plaza Hotel & Casino, in Atlantic City, New Jersey, has been unveiled to enthusiastic acclaim by guests and travel industry professionals. The extensive facelift, designed by WESTAR Architects, updates the Casino, Liquid Bar, 24 Central Café, Baccarat Pit and Asian Gaming areas and facilities collectively encompassing some 75,000 square feet of space, to reveal a world guests are unlikely to see anywhere else. Guests are drawn into the new environments—all lavishly adorned in onyx, rosewood, crystal, gold leaf, Venetian plaster and carpet, accented by spectacular chandeliers and marble mosaics, and saturated in a color palette of red, blue, beige, black and gold—by following pathways that offer intriguing vistas and focal points. This layering of successive imagery has the effect of enticing guests to discover what seems tantalizingly close yet always slightly beyond reach, a phenomenon that instantly beguiles anyone in the Crystal River area of the Casino. Here, the impact of a ceiling of multiple chandeliers that is reflected in the marble tile mosaic floor directly below, with even lighting over the gaming activity and sparkle over the pathway, regularly elicits a "wow" from guests. entrance.

Clockwise from far left: Baccarat Pit; bar lounge; Asian Gaming; casino.

Photography: Darius Kuzmicka.

WESTAR Architects

Hannah's Neighborhood Bistro
Las Vegas, Nevada

Water commands attention in the desert, and clearly fascinates guests at the popular, new 245-seat, 7,200-square-foot Hannah's Neighborhood Bistro, in Las Vegas, designed by WESTAR Architects. In a once anonymous commercial structure remodeled by restaurateur Hannah An to serve Asian and French cuisine inspired by her family's Vietnamese heritage, an informal interior resembling a French colonial home now prevails. Three water features set the stage for bar and lounge, formal and casual dining rooms, sushi bar and outdoor patio with satay bar and grill. Guests view a small waterfall and "Walk on Water" glazed koi pond recessed into the floor—a signature of the An family's Crustacean Restaurant in Beverly Hills—upon entering, and gaze in awe at the sushi bar's 30-foot-long waterfall. Hannah's offers more than water, of course. Its residential ambiance, crafted with wood floors, river rock aisles, Venetian plaster walls, dramatic lighting, ceiling fans, and resort-style furniture, is accented with Asian art and silk draperies that introduce a measure of visual separation, with each space taking on a character of its own. As guests happily agree, it's an appetizing setting for such entrées as shaken beef, crab fried-rice, and garlic noodles-an An family favorite.

Top left: Dining room.
Top right: Exterior.
Left: Outdoor patio.
Bottom left: Bar.
Photography: Darius Kuzmickas.

Worth Group, Architects & Interiors, PC

Reno
9400 Gateway Drive
Suite B
Reno, NV 89521
775.852.3977
775.852.6543 (Fax)
www.worthgroup.com

Denver
384 Inverness Parkway
Suite 100
Englewood, CO 80112
303.649.1095
303.649.1098 (Fax)

Vegas
5888 West Sunset Road
Suite 201
Las Vegas, NV 89118
702.869.9354
702.869.9899 (Fax)

Worth Group, Architects & Interiors, PC and G.S. Ridgway & Associates, Inc.

French Lick Resort Casino
French Lick, Indiana and West Baden, Indiana

For over two centuries, the mineral springs of French Lick, Indiana have drawn a remarkable stream of guests, including such celebrities as financier "Diamond Jim" Brady, gangster Al Capone, composer Irving Berlin, actress Lana Turner and President Franklin D. Roosevelt. Today, the springs are working their magic again with the help of the new 689-room, one million-square-foot-plus French Lick Resort Casino. By combining the restoration of two hotels listed on the National Register of Historic Places, the 443-room French Lick Springs Hotel, in French Lick (whose existing structure dates from 1901), and the 246-room West Baden Springs Hotel, in West Baden (whose existing structure dates from 1902), with the construction of the new, 42,000-square-foot French Lick Casino, in French Lick—all designed by WorthGroup Architects & Interiors and G.S. Ridgway—French Lick emerges as a unique property that returns gaming to southern Indiana for the first time since 1949 and helps to revive the region's economy. Since the restoration seamlessly blends historic form with contemporary function, today's guests can experience many of the leisure activities known to previous generations along with more contemporary ones. Even without the casino, French Lick's two hotels (located one mile apart and connected by shuttle buses) offer numerous possibilities in their reception areas, lounges, 17 food and beverage venues (including fine dining restaurants, pub restaurants, buffet, Casino bistro, and more), kitchens, events center, bowling alley with arcade, indoor and outdoor pools, two spas, two golf courses, retail shops, parking structure, and extensive support areas. Of course, the creation of an ambitious, historically

Top right: French Lick lobby.

Above middle: West Baden Fine Dining Restaurant.

Above: West Baden King Guestroom.

Left: Entrance to West Baden.

Opposite: West Baden Atrium.

Photography: Bob Perzel.

Worth Group, Architects & Interiors, PC and G.S. Ridgway & Associates, Inc.

accurate and guest-friendly facility imposed numerous challenges. Among the more noteworthy were the National Park Service's required approval of the renovation, the Indiana Gaming Commission's mandated order for a casino resembling a classic riverboat (floating in its own moat for legal purposes) to begin operations within 18 months of the award of the gaming license, the need for asbestos abatement, the custom design of furnishings for guestrooms that are generally smaller in floor area and height than contemporary standards, the use of earthwork to raise the elevations at selected portions of the site, which occupies a floodway, the reorganization of spaces and circulation within the historic structures for new uses and better wayfinding without violating their authenticity, and the choice of colors, materials, furnishings and lighting to support the facility's distinctive and memorable guest experience—available, like the mineral springs, only in French Lick.

Above left: French Lick casino.

Above right: French Lick fine dining.

Right: West Baden lobby.

Below left: West Baden spa locker room.

Below right: West Baden fine dining wine tasting.

EXIT

Worth Group, Architects & Interiors, PC

Inn of the Mountain Gods Resort & Casino
Mescalero, New Mexico

You aren't dreaming when you awaken to clean mountain air, breathtaking scenery and exciting entertainment options at the Inn of the Mountain Gods, perched among the picturesque mountains of Mescalero, New Mexico. It only seems that way, thanks to the Inn's award-winning, 277-room, 1.1 million-square-foot replacement facility, designed by WorthGroup Architects & Interiors following a

thorough reappraisal of existing site/master planning, architecture, interior design and lighting design. The revitalized Inn now blends into its sloping site rather than stand out from it, so its five-story, 1,200-foot-long façade appears only two stories high from the roadside. In addition, the new design consolidates gaming, meeting and hotel functions under one roof instead of isolating them in separate structures, enriches its outdoor and indoor settings using cultural themes from the Inn's owner, the Mescalero Apache Tribe, without being overpowering, and supports the Inn's many outdoor activities through a contemporary interior design that updates the traditional mountain ski lodge while providing generous scenic views of the grounds. A renewed sense of discovery is evident from the entry's rotunda porte-cochère, reception area and reception lounges to the Grand Hall and the major destinations beyond it, including the Casino and its Big Game sports bar, poker room, 40,000 square feet of convention space, fine dining restaurant, piano

Top: Exterior on Lake Mescalero.

Above left: Wendell's Steak and Seafood Restaurant.

Above right: Wendell's Lounge.

Left: Grand lobby.

Photography: John Wong Photography (top, left); Glenn Cormier (above left and above right).

Left: Casino.

Lower left: Guest suite living room.

Photography: John Wong Photography, Glenn Cormier.

lounge, buffet, nightclub, pool, fitness center and guestroom accommodations, as well as Ski Apache and other outdoor activities. Such welcome features as the radial floor plan that fans out from the entry, the bold, contemporary architecture that complements the surrounding mountains, the palette of natural tones with punches of color, timeless materials like stone, wood and metal, and comfortable furnishings with a lodge-like feel explain why the Inn's CEO, Brian Parrish, has declared, "People are blown away by the resort, which far exceeded our expectation."

You're Going to Love This

How do you design hotels and restaurants where guests feel comfortable enough to try something new–and have great experiences they will remember long after?

By Roger Yee

Who hasn't felt like the physician Dr. Otternschlag at the end of the classic 1932 film, The Grand Hotel, set in Berlin's opulent, Art Deco-style Grand Hotel? Having failed to notice all the drama around him or receive any messages at the front desk, he closed the film with the famous lines: "The Grand Hotel. Always the same. People come. People go... nothing ever happens." Fortunately, guests visiting today's hotels and restaurants find themselves being actively courted to win their current business and long-term loyalty, with no effort spared to entice them into trying their amenities. If anything, the competition for customers is intensifying. Along with a surge in business and leisure travel worldwide—enabling the U.S. lodging industry to record annual sales of $133.4 billion (2006) and the U.S. restaurant industry to report annual sales of $511.4 billion (2006)—the hospitality industry is actively consolidating existing businesses, testing new concepts, and expanding into developing markets. This burst of activity is challenging architects and interior designers to respond to a widening range of locations, cultures and markets. Consider these examples.

• Major international hotels opening in Beijing, such as the Ritz-Carlton Beijing, must diligently cultivate and coordinate local suppliers and design institutes to source the bulk of their building materials, M/E/P systems and interior furnishings if they want to offer world-class accommodations in the capital of the nation that will soon overtake Germany as the world's third-largest economy.

• Chicago's Republic Restaurant is the latest venture by husband-and-wife restaurateurs from Shanghai, who are moving upscale from popular neighborhood Chinese establishments to a hip, Pan Asian place in the trendy River North district, and filling their stylish, minimal environment with hipsters and "suits" alike.

• What began simply as Bliss, a New York City spa celebrated for its fun, "no attitude" atmosphere, is now embarking on a global expansion, sustained not only by its exceptional services and popular skin care and home spa products, but also by Starwood Hotels & Resort, which acquired it to develop new units wherever youthful, stylish and affluent populations convene in W Hotels around the world.

Which key design trends represent what hoteliers and restaurateurs are typically doing to increase the effectiveness of their spaces? While the replies from leading architecture and interior design firms do not disclose startling revelations, they reveal how much the global economy is drawing distant outposts of the

Above: This spectacular bath in the Presidential Suite of the new, 235-room Swissôtel Krasnye Holmy,designed by BBG-BBGM, appeals directly to business travelers seeking modern, world-class luxury in Moscow.

developing world into the commercial and cultural life of the industrialized nation–and vice versa. Among the topics designers cite are branding, guest experiences, green design, and ultra-luxury service and its opposite, no-frills service.

Can space be branded and not become boring?

Popular hotels and restaurants have well developed characteristics in the eyes of their owners and operators as well as their customers, just as unpopular ones have. But the consolidation of the industry in recent years has forced the industry to take a more methodical look at the question of branding, the establishment and promotion of an organization's identity. "The rise of Starwood Hotels & Resorts Worldwide was a defining moment in the branding of the hospitality industry," observes Julia Monk, AIA, IIDA, ASID, LEED AP, a partner of BBG-BBGM. "The company found itself with properties shoulder to shoulder that were too similar, and needed to create distinctions among them." As a result, she adds, W would become "hip," Hilton "classic" and Sheraton "family oriented."

Meaningful differentiation among hotel and restaurant groups has not only enabled guests to make appropriate choices, it has shown the hospitality industry how to keep introducing new brand concepts. (According to PricewaterhouseCoopers, 34 U.S. hotel brands were launched between 2005 and 2007.) As a result,

many owners and operators pay serious attention to the branding of their facilities, occasionally retaining brand consultants to meet their designers before they create facilities with a branded, signature look. Even top chefs such as Tom Colicchio, Wolfgang Puck and Thomas Keller are laboring over their growing empires of dining rooms and exhibition kitchens nearly as much as their cuisine and service.

"Branding can cover an entire resort or restaurant, meaning a collective vision of everything the public views," explains Richard Brayton, FAIA, a principal of BraytonHughes Design Studios. Careful cultivation of that vision by client and designer explains why a guest can easily recognize he or she is in a Four Seasons hotel or Ruth's Chris steakhouse anywhere in the world. "At an individual level," Brayton continues, "branding can represent the way to fulfill a customer's expectations." An urban hotel hoping to keep a business guest playing over the weekend after spending the week working, he suggests, might want the environment to explicitly showcase the presence of first-class restaurants and a spa as well as good meeting rooms.

Above: A classic, modern look infused with traditional northern Chinese motifs gives the 500-room Regent Beijing its own unique style as major international hotels rush to establish a presence in Beijing.

Yet branding the physical environment is neither a panacea nor an immutable standard for success. "Branding starts with functional issues, of course," states Howard Wolff, senior vice president of WATG. "You don't want to risk having poor operations that will negate the value of novel concepts. At the same time, branding must change with the times and the context." Though Ritz-Carlton once felt that being "ladies and gentlemen serving ladies and gentlemen" meant addressing you as "madame" or "sir" and escorting you to the front desk, Wolff observes, it now acknowledges that many guests do not want this. He also points out that the hotel's archetype interior design, based on an historic 18th-century English style rooted in the culture of upper class Northeasterners, has discreetly yielded to local influences. Thanks to the rising comfort level of sophisticated guests with exotic cultures and cuisines, the Ritz-Carlton in Laguna Niguel will not be mistaken for its counterpart in Boston.

The customer is always right: How guest experience, green design, and other trends are shaping hospitality settings

Interestingly, the influence of indigenous culture on branded environments may be welcomed yet limited in developing nations such as Russia, China and India. American and other Western designers working abroad have discovered that their enthusiasm to absorb and use the design vocabularies of host nations in their work may be tempered by the desire of local owners and operators to project an image that breaks with entrenched standards and traditions. "Overseas clients bring us on board for our modern and international approach to planning and design," reports Johnny Marsh, an associate of tonychi and associates. "They may want us to absorb local cultural influences in their facilities, but they hire us to develop modern facilities that meet the latest international standards. In fact, they want their guests to know this."

Designers asked by hoteliers and restaurateurs to create settings for unique guest experiences say form and function can combine in imaginative ways to turn moments spent at the lobby and registration desk, guest bedroom and bath, restaurant, fitness center or spa into satisfying and even memorable occasions. Yet they add that design is just one variable among many that a guest senses. "Design alone won't make a guest experience unique and satisfying," declares Miguel Baeza, a principal of HBA/Hirsch Bedner Associates. "Service also provides critical sources of differentiation."

Baeza acknowledges that designers enjoy every opportunity to extend the conceptual and physical boundaries of space, such as the growing guest bedrooms and baths in high-end hotels, now approaching some 600 to 800 square feet per guestroom unit. "Designing these guestrooms is like designing small apartments," he indicates. Yet the operating details will also count, with the quality of room service, bed linens, communication and entertainment systems, bath towels, hair drier and toiletries saying as much as the facility does about who a hotelier or restaurateur is and how he treats his guests.

Above: Preferred by countless celebrities and heads of state, the Intercontinental Genève, in Geneva, Switzerland, is being updated to produce a completely new, classically modern look, keeping pace with its sophisticated clientele.

Green design, which generally means designing for sustainability, is steadily gaining adherents in hospitality. "Green design has become a top priority among hotels, with some even seeking LEED accreditation," declares Marsh of tonychi and associates. "From design and construction to operations, new opportunities for eco-friendly solutions are being found." Architecture more readily lent itself to environmentally sound design solutions, but interior design is quickly catching up with the support of guests and employees. As Marsh illustrates, a large furniture order for a major hotel or restaurant can effect how wood is grown and harvested, materials are selected, and finished products are manufactured, introducing a trickle-down effect that can benefit other and smaller projects.

As for the hospitality industry's continued testing of markets at the high and low ends of the market, designers sound eager to rise to the challenge of creating appropriate new physical environments for them. "Life at the top keeps rising," BBG-BBGM's Monk says. "The world has some eight million millionaires and one thousand billionaires. So the bar keeps rising, and we hear about seven-star service. Fortunately, there may be a trickle down effect in the process, bringing what were luxury amenities like spas to lower levels of service and benefiting more customers." She notes that design is even helping low-cost hotels such as Courtyard by Marriott and quick-service restaurants like McDonald's to build customer loyalty.

In the end, good design for hospitality remains firmly rooted in knowing who is being served. "Long before branding and guest experience became common terms, our practice spent a lot of time at the front end of projects defining what we call the customer 'DNA' or differentiators, nuances and attitudes," recalls Mark Knauer, a principal of Knauer Inc. "The answers help generate not just the design and graphics, but also the food and service. You can take this DNA, like the seeds of an apple, anywhere."

Knauer admits the process is far from easy. "Defining your differentiators is half the battle," he finds. "Design is relatively easy after you've determined the big 'D.'" But the rewards are potentially enormous, he argues. When a hotel or restaurant can honestly and accurately respond to the basic issues of "Who are we?" and "Who does the customer think we are?" it can devise effective strategies to combine design and service into one or more successful businesses.

"The public is always on a pendulum going from a time for comfort to a time for experimentation and back," Knauer concludes. "There will always be room for newcomers with new ideas." In other words, a new hotel or restaurant concept is probably waiting in the wings every time we ask, "Where shall we stay?" or "What's for dinner?"

Above: In this view of the main dining room, Entourage, a 475-seat steakhouse in Schaumburg, Illinois, takes America's love affair with steak inside a sassy, Retro tribute to the 1960s world of Frank Sinatra's Rat Pack.

Resources*

Atlantis, The Cove, Atlantis
Design Firm: WATG
Furniture: Blue Leaf Hospitality

Atlantis, The Residences at Atlantis
Design Firm: WATG
Furniture: Blue Leaf Hospitality

Beau Rivage High Limit Gaming
Design Firm: Cleo Design
Furniture: Baker Knapp & Tubbs, Chair Choice, Lily Jack, Spectrum, Ltd.
Carpets & Flooring: Couristan
Fabrics: Ashbury Hides, Jim Thompson, Myung Jin Mohair, Tiger Imports
Lighting: Spectrum, Ltd.
Wallcoverings and Paint: Larsen
General Contractors: Ruzika Company

Caesar's Palace Las Vegas, Concierge Suite
Design Firm: HBA/Hirsch Bedner Associates
Furniture: A. Rudin, Asher Cole, Bolier & Co., Gerard, Giorgetti, Jan Showers, Lewis Mittman, Lockhart Collection, NFK International, Victoria Hagan
Carpets & Flooring: Aqua Hospitality, Saroyan, Shaw Carpet
Fabrics: Bergamo, Dedar, Designtex, Innovations, Moor & Giles, Old World Weavers, Pindler & Pindler, P. Kaufmann, Prismatek, Travers, Valley Forge Fabrics, Yoma Innovations
Lighting: Aqua Creations, Charles Lamp, City Studio, Jan Showers, Lumis srl, Moones, Orestes Suarez, Porta Romana, Remington Lighting, Zia-Priven
Wallcoverings and Paint: Art People, Dunn Edwards, Innovations, Maya Romanoff, Metro Wallcoverings/Sanitas
Lighting Consultants: Elwyn Gee

California Pizza Kitchen Foxwoods, Foxwoods Resort & Casino
Design Firm: Aria Group Architects, Inc.
Furniture: Commercial Custom Seating
Carpets & Flooring: DalTile, DSF Flooring
Fabrics: Valley Forge
Lighting: Capaul
Wallcoverings and Paint: Sherwin-Williams, Wolf Gordon
Window Treatments: Levolor
General Contractors: Shawmut Design & Construction
Lighting Consultants: Mark Schulkamp

Cirque du Soleil, The Beatles "Love" Cirque du Soleil Theater
Design Firm: Marnell Architecture
Furniture: Irwin Seating
Carpets & Flooring: Corradini, Crossville, DalTile, Ulster Carpets
Fabrics: Maharam, Momentum
Lighting: Omni Lighting, Triton Chandelier
Wallcoverings and Paint: Hospitality Network, Metro Wallcovering, Sanitas
General Contractors: Marnell Corrao Associates

Cliff House
Design Firm: BraytonHughes Design Studios
Furniture: Blue Leaf Hospitality
Lighting: Frank Neidhardt Chandeliers
Window Treatments: David Verna, Peninsulators
Furniture: Blue Leaf Hospitality
General Contractors: Nibbi Brothers
Lighting Consultants: Auerback + Glasow

COEX Seven Luck Casino
Design Firm: LBL Architecture & Interiors
Furniture: Poongjin Furniture
Carpets & Flooring: Couristan, Grama
Fabrics: Archiline, Hyunwoo Design
Lighting: Color Kinetics
Ceilings: Poongjin Furniture
Wallcoverings and Paint: Archiline
Window Treatments: Hyunwoo Design
General Contractors: Poongjin IO
Lighting Consultants: John Levy Lighting

Fairmont Heritage Place, Acapulco Diamonte
Design Firm: sfa design
Fabrics: Malabar
Kitchen & Bath Fixtures: Kohler

Fairmont Turnberry Isle Resort & Club
Design Firm: sfa design
Furniture: Fairmont Designs, Holly Hunt
Carpets & Flooring: Clayton Miller, Walker Zanger
Fabrics: JLF Contract Furnishings, 3form
Lighting: Belvedere Artisan, Hallmark
Wallcoverings and Paint: Majestic Mirror & Frame, Rosenbaum Fine Art

Gansevoort Hotel
Design Firm: Stephen B. Jacobs Group, Andi Pepper Interior Design
Furniture: American Atelier, Bright, Charter Furniture, HBF, JL Furnishings, Murray's Ironworks, Zographos Designs
Carpets & Flooring: Atlas, Couristan
Fabrics: Bramson House, Donghia, Kravet
Lighting: Hallmark Lighting, International Ironworks, Lee's Studio, Resolute, Ruth Drachler Co.
Wallcoverings and Paint: Wolf Gordon
Window Treatments: Bramson House, Kravet, Lee Jofa Fabrics
General Contractors: Levine Builders

Griswold Inn Wine Bar
Design Firm: Haverson Architecture and Design
Furniture: Action! Marketing, New York Custom Interior Woodworking Corp.
Carpets & Flooring: Carlistle Restoration Lumber Inc.
Fabrics: ArcCom, Decorative Vinyland Fabric Co.
Lighting: Lightolier, Urban Archaeology
Ceilings: Benjamin Moore, New York Custom Interior Woodworking Corp.
Wallcoverings and Paint: Benjamin Moore, Duralee, Litex Finishing Systems, Maharam
Window Treatments: Nanik
General Contractors: Paul Riggio Contracting
Lighting Consultants: Haverson Architecture

Harrah's New Orleans Hotel
Design Firm: Marnell Architecture
Furniture: Newman Frey, Westwood Interiors, William Switzer
Carpets & Flooring: American Tile Co., Zenith International
Fabrics: Beacon Hill, Leslie Hannon, Metaphores, Old World Weavers, Osborne & Little, Randolph & Hein, Schumacher
Lighting: Lurid, Murray's Ironworks, Nickmair Weeks, Triton Chandeliers
Wallcoverings and Paint: ArcCom, Akleth Delevee, MDC Wallcovering, Studio E, Trik-kes Wallcovering
Window Treatments: Coast Drapery
General Contractors: NOW2, LLC

Hilton Grand Vacations Club on the Las Vegas Strip
Design Firm: MBH Architects
Furniture: A. Rudin Designs, Decca Hospitality
Carpets & Flooring: Atlas, Couristtan, Crossville Tile, Daltile, Shaw, Walker Zanger
Fabrics: Carnegie, Knoll Textiles, Pindler & Pindler, Pollack & Assoc., Unika Vaev
Lighting: Illuminating Experiences, Minka

**An Incomplete list of major sources.*
For more information please call design firms.

Lavery, Orion Chandeliers, Sirmos
Ceilings: Armstrong
Wallcoverings and Paint: JM Lynne, Koroseal, Lentex

Hokusai
Design Firm: LBL Architecture & Interiors
Furniture: Aceray, Plaxis
Carpets & Flooring: Junckers, Longust Ceramic Tile
Fabrics: Cort Na Leathers, Kravet, Valley Forge
Lighting: Artemide, Elite Ideas
Ceilings: Dunn Edwards
Wallcoverings and Paint: Ann Sacks, Dunn Edwards, 3-Form, Wolf Gordon
General Contractors: RTM Construction Services, Inc.
Lighting Consultants: Integrated Lighting Design

Hotel Giraffe
Design Firm: Stephen B. Jacobs Group, Andi Pepper Interior Design
Furniture: Beverly, Breuton, HBF, JLF, Lewis Alan, Mike Moore Portfolio, Niedermaier
Carpets & Flooring: Andi Pepper, fabricated by LL Floor Design and Helios, Washington Square
Fabrics: Duralee, Hines, Glant, Holly Hunt, Kravet, Melrose House, Rose Tarlow
Lighting: American Glass Light Co., Boyd
Wallcoverings and Paint: Maya Romanoff, Silk Dynasty Inc.
Window Treatments: Kravet
General Contractors: HRH Construction
Lighting Consultants: Horton Lees Lighting

Hotel Solamar
Design Firm: Intra-Spec Hospitality Design, Inc.
Furniture: A. Rudin Furniture, Century Furniture, Decca Hospitality Furnishings, Fong Brothers, HBF, Holly Hunt, Keilhauer, Kettal, Royal Custom Designs, Walters Wicker
Carpets & Flooring: Decorative Carpets, Durkan, Pacific Coast Mat, Ulster Carpets
Fabrics: Architex, Clarence House, Donghia, Edelman Leather, Great Plains, Hinson & Co., Jim Thompson, Maharam, Malabar, Osborne & Little, Perennials, P. Kaufmann, Ralph Lauren Home, Sunbrella
Lighting: Aqua Creations, Burt's Cason, Fanimation, Global Lighting, La Spec, Lumiere, Murray's Ironworks, Sirmos
Wallcoverings and Paint: Metro Wallcoverings, Sellers & Josephson
Window Treatments: Danmer Custom Shutters, 5-Star Interior Services
General Contractors: Davis Reed Construction
Lighting Consultants: Terry Ohm

Hyatt Regency McCormack Place, Forno Cafe, Shor Restaurant, M/X Lounge, Board Room
Design Firm: Aria Group Architects, Inc.
Furniture: Ahdreu World, Bernhardt, Environetics, Falcon, Gar, HBF, Lily Jack, Sandler
Carpets & Flooring: Atlas, Azea, Bentley Prince Street, Imola Ceramiche, Shaw
Fabrics: ArcCom, CF Stinson, HBF, Knoll, Luna, Maharam, Momentum, Pollack
Lighting: Challenger, Cooper, Doyle Signs, Juno
Ceilings: USG
Wallcoverings and Paint: Benjamin Moore, MDC Wallcoverings, Modular Arts
Window Treatments: Skyline Design
General Contractors: Crown Construction
Lighting Consultants: Schuler Shook

Hyatt Regency, Newport, RI
Design Firm: DiLeonardo International, Inc.
Furniture: American of Martinsville, Haworth, Marquis Custom Hospitality Seating
Carpets & Flooring: Atlas
Fabrics: ArcCom, P. Kaufmann, Valley Forge Fabrics
Lighting: Challenger Lighting, 100 Watt Network, Wilshire Lighting
Wallcoverings and Paint: D.L. Couch, Valley Forge Fabrics
Window Treatments: Knoll
Lighting Consultants: D. Schweppe Design

InterContinental, Lagos
Design Firm: DiLeonardo International, Inc. (Dubai Office)
Furniture: A. Rudin Furniture & Textiles, Bernhardt, Daniel Paul, Davis, Georgetti USA, Henry Nall Design, Mitylite, Royal Custom Designs, Vanguard Furniture
Carpets & Flooring: Brintons Carpet, Shaw Carpet
Fabrics: Beacon Hill, Cortina Leather, Daniel C. Duross, Decorator's Walk, Edelman Leather, Elan Textiles, Kravet, Markasia Ltd., P. Kaufmann, Sina Pearson, Valley Forge
Lighting: Abensal Lighting, Fer House Lighting, McGuirre Furniture Co., Quasar, Tijan
Wallcoverings and Paint: Globe Coat & Jotlin Paint, W/C by Vescom
Window Treatments: Hunter Douglas
Lighting Consultants: Project Lighting Design

JIA Restaurant
Design Firm: Cleo Design
Furniture: Kisabeth Furniture, Mark David, Troy Wesnidge
Carpets & Flooring: Smith and Fong, Terrazzo
Fabrics: Donghia, HBF Textiles, Illuminaire, Innovations Textiles, Kravet, Mokum Textiles
Lighting: Orion Chandeliers
Water Feature: BluWorld USA
Window Treatments: Coast Drapery
Lighting Consultants: Ruzika Company

Jumeirah Carlton Tower, Gilt Champagne Bar
Design Firm: HBA/Hirsch Bedner Associates
Furniture: Nicky Brady Upholstery
Carpets & Flooring: EE Smith, Steve Charles Metal Brass Tile
Lighting: Suzan Etkins
Wallcoverings and Paint: Donghia, Muirhead

Junior's
Design Firm: Haverson Architecture and Design
Furniture: JC Furniture, Nevamar, New England Surfaces
Carpets & Flooring: AKDO
Fabrics: Knoll Textiles, Maharam
Lighting: Apex Lighting, Bocci, Rejuvenation
Ceilings: Benjamin Moore
Wallcoverings and Paint: D.L. Couch, Glacier Glass Tile, Stone Source
General Contractors: Maristech Construction Corp.
Lighting Consultants: Haverson Architecture and Design

Lake Arrowhead Resort and Spa
Design Firm: Intra-Spec Hospitality Design, Inc.
Furniture: Decca Hospitality Furnishings, Drexel Heritage, Ferguson Copeland, Global Views, Janus et Cie, Kashwere, Lily Jack, Royal Custom Design, Summit Furniture
Carpets & Flooring: Durkan, Smith & Fong
Fabrics: Donghia, Edelman Leather, HBF Textiles, Hinson & Co., Kravet, Moore & Giles, Perennials Outdoor Fabrics, P. Kaufmann, Quadrille, Robert Allen, Summer Hill, Townsend Leather, Valley Forge Fabrics
Lighting: Fine Art Lamps, Flambeau, La Spec, Murrays Ironworks, Stonegate Designs
Wallcoverings and Paint: Benjamin Moore, Dunn Edwards, JM Lynne, Sellers & Josephson
Window Treatments: Danmer Custom Shutters, 5-Star Interior Services
General Contractors: Precision Hotel Interiors

Library Hotel Expansion
Design Firm: Stephen B. Jacobs Group, Andi Pepper Interior Design
Furniture: American Atelier, Bernhaardt, Bright, Charter Furniture, Cortina Leather, Hallmark/Ruth Drachler, HBF, JL Furnishings, Orlando Diaz-Azcuy Collection, Sandler/Sam Gordon Associates, Triconfort, Walters Wicker, Zographos Design
Carpets & Flooring: Atlas, Couristan, Patterson, Flynn & Martin, Stone Source
Fabrics: Donghia, Kravet, Yoma
Lighting: Bruck, International Ironworks, Lee's Studio, Resolute
Wallcoverings and Paint: Maya Romanoff, Wolf Gordon
Window Treatments: Bramson House
General Contractors: Levine Builders
Lighting Consultants: Horton Lees

The Mandarin Oriental Hotel
Design Firm: Intra-Spec Hospitality Design, Inc.
Furniture: Baker Furniture, Decca Hospitality Furnishings, Henredon, Kay Chesterfield, Lily Jack, Royal Custom Designs, Wood World
Carpets & Flooring: Brintons, Durkan, Ulster Carpets
Fabrics: Architex, Bergamo, Kravet, P. Kaufmann, Quadrille, Valley Forge Fabrics, Zax
Lighting: Baker Furniture, Chapman Lighting, Jamie Young, La Spec
Wallcoverings and Paint: Benjamin Moore, Metro Wallcoverings
Window Treatments: 5-Star Interior Services, Schwimmer Drapery

Market Bar
Design Firm: BraytonHughes Design Studios
Carpets & Flooring: ASN Natural Stone, Golden State Flooring
Fabrics: Donghia
Lighting: Omega Too Lighting
Ceilings: San Francisco Victoriana, WF Norman
Window Treatments: Franciscan Interiors
General Contractors: Plant Construction
Lighting Consultants: BraytonHughes Design Studios

MGM Grand, MGM Race and Sports Book, Poker Room, and Centrifuge Bar
Design Firm: Steelman Partners
Furniture: Eric Brand Furniture, Gasser
Carpets & Flooring: Ulster
Fabrics: Knoll, Maharam, Momentum, Pollack and Associates
Lighting: Lightology, Niedermaier
Ceilings: Whisper Wall
Wallcoverings and Paint: JM Lynne, Koroseal
Window Treatments: Coast Drapery
General Contractors: Perini and Madison
Lighting Consultants: shop12

Mirabel
Design Firm: BraytonHughes Design Studios
Furniture: Archittectural Woodwork of Montana, National Upholstering, West Coast Industries
Carpets & Flooring: Brintons
Fabrics: Lou Peacock & Associates, Marvic Textiles, Moore & Giles
Lighting: Enid Ford, International Ironworks
Wallcoverings and Paint: Benjamin Moore
General Contractors: Linthicum Custom Builders
Lighting Consultants: Bouyea & Associates

Mohegan Sun Hotel & Casino, Uncas American Indian Grill and Chef's Bagels
Design Firm: Haverson Architecture and Design
Furniture: Action! Marketing, Beaufurn, Henry Calvin, ISA International, Table Topics
Carpets & Flooring: Ann Sacks, Atlas, Karnadean Vinyl Wood
Fabrics: Interlam, Liz Jordan-Hill, Maharam, Maya Romanoff, Momentum, Robert Allen, 3-Form
Lighting: Color Kinetics, Lightolier
Ceilings: Sherwin-Williams
Wallcoverings and Paint: Benjamin Moore, Donghia, Kokomo Glass, Rudy Art Glass, Sherwin-Williams, Yangki
General Contractors: Mohegan Sun Tribal Development
Lighting Consultants: Focus Lighting

The Muse Hotel
Design Firm: Intra-Spec Hospitality Design, Inc.
Furniture: Bernhardt, CSA, CSN Office Furniture, Global Views, HBF, Janus et Cie, Keilhauer, Royal Global Designs, William Montague
Carpets & Flooring: Couristan, Durkan, Gammapar, Legno Veneto
Fabrics: Edelman Leather, Kravet, P. Kaufmann, Polllack & Associates, Valley Forge Fabrics
Lighting: Boyd Lighting, Hallmark Lighting
Wallcoverings and Paint: Benjamin Moore, Designtex, Koroseal, Wolf Gordon
Window Treatments: Hunter Douglas
General Contractors: Faithful & Gould, R.P. Brennan

Oak Valley Ski Resort
Design Firm: LBL Architecture & Interiors
Furniture: Gio International, Kesson International
Carpets & Flooring: Cosis International, Flora Hardwood Flooring
Lighting: Dong Myung Lighting, Jin Young Lighting
Ceilings: Wattyl Australia
Wallcoverings and Paint: Fidelity, RJF International, Sirpi, Wattyl Australia
General Contractors: Kesson International

Oakwood Premier COEX Center, Penthouse Presidential Suite
Design Firm: LBL Architecture & Interiors
Furniture: Heehoon Art Furniture
Carpets & Flooring: Dome Design, Grama
Fabrics: Archiline, Hyunwoo
Lighting: Colux, Kukje
Ceilings: Heehoon
Wallcoverings and Paint: Archiline
Window Treatments: Hyunwoo Design
General Contractors: Heehoon D&G
Lighting Consultants: Colux Lighting, Kukje Lighting

Okada Restaurant, Wynn Resort Las Vegas
Design Firm: HBA/Hirsch Bedner Associates
Furniture: Ashbury Hides, Fong Brothers
Carpets & Flooring: Emma Gardner Design, The Natural Carpet Co., Stone Connection
Fabrics: Bergamo, Joseph Noble Co. c/o Dakota Jackson
Lighting: Aqua Creations, Triton Chandelier

Pan Pacific Seattle
Design Firm: HBA/Hirsch Bedner Associates
Furniture: Cheng Meng Furniture
Carpets & Flooring: Couristan
Fabrics: Majilite, Valley Forge
Lighting: Baldinger, Hallmark, Unilight
Wallcoverings and Paint: Carolyn Ray Inc., Designtex, MDC Wallcoverings, Metro Wallcoverings, Tri-Kes Wallcovering, Wolf Gordon

Parc Rittenhouse
Design Firm: Daroff Design
Furniture: All Wood Treasures, Baker, Kravet
Carpets & Flooring: Junckers, Mark Phillips, Stone Source
Fabrics: Architex, Fabricut, Garrett Leather, Jacques Bouvet et Cie, Spinneybeck
Lighting: Boyd Lighting, J. Harris & Sons, Sistemalux

Wallcoverings and Paint: Benjamin Moore
General Contractors: The Sullivan Company

Radisson Medeu Resort, Kazakhstan
Design Firm: DiLeonardo International, Inc.
Furniture: Century Furniture, Donghia, HBF, Maura Starr, Royal Custom
Carpets & Flooring: Markasia
Fabrics: Daniel Duross Leather, Edelman Leather, Knoll, Rodio Fabrics, Sina Pearson
Lighting: Chelsom, Flux Lighting, ITRE Lighting, Quasar Lighting, Zaneen
Wallcoverings and Paint: Designtex, Koroseal, Maya Romanoff, MDC, Wolf Gordon
Lighting Consultants: PHA Lighting Design

Rae
Design Firm: DAS Architects, Inc.
Furniture: General Seating Solutions, L&B Industries, Wood Goods
Carpets & Flooring: Bizazza Glass Tile, Casa Dolce Ceramic Tile, Constantine
Fabrics: Anzea, Designtex, Maharam, Pallas, Pollack
Lighting: Architectural Details, Cooper, Floss, Inter Luce
Wallcoverings and Paint: Carnegie, HBF Textiles, Knoll, Sina Pearson
Window Treatments: Lister Upholstery & Window Fashions
General Contractors: Domus
Lighting Consultants: The Lighting Practice

Riche Restaurant
Design Firm: Marnell Architecture
Furniture: Charter Furniture, L&B Furniture, Mark David, Thomas & Associates, Vaughan Benz, Westwood Interiors
Carpets & Flooring: Carlisle Restoration Lumber, Scott Group Carpets
Fabrics: Architex, Crezana Design, Fabricut, Majilite, Nobilis, Ralph Lauren, Robert Allen
Lighting: City Studio, Holly Hunt Lighting, Nowell's Inc., Palmer Hargrave, SchonbeckTriton
Wallcoverings and Paint: Akleth Delevee, David Goldberg Design
Window Treatments: Coast Drapery
General Contractors: NOW2, LLC

The Rittenhouse Hotel
Design Firm: Daroff Design
Furniture: Baker, HBF
Carpets & Flooring: Brintons
Fabrics: ArcCom, Brunschwig & Fils, Fabricut, Kravet, Penaz, Stroheim & Roman
Lighting: J. Harris & Sons, Kichler
Wallcoverings and Paint: Benjamin Moore, MDC Wallcovering
Window Treatments: Brunschwig & Fils, Kravet

Ritz-Carlton Beijing, Financial Street
Design Firm: HBA/Hirsch Bedner Associates
Furniture: Strong CASA
Carpets & Flooring: Kingtop
Lighting: Lightsource International
General Contractors: Hong Gao, Guangzhou Meishu, Shanghai Xin Li, Shengzhen Changchen

Rosa Mexicano
Design Firm: Mancini•Duffy
Furniture: Quality Seating
Carpets & Flooring: DalTile
Fabrics: Valley Forge Fabrics
Wallcoverings and Paint: Daisy Cake, Fluent Design, Habitus
Window Treatments: Image King
General Contractors: Shakman Hospitality
Lighting Consultants: Cosentini Lighting Design

Salt Creek Grille
Design Firm: DAS Architects, Inc.
Furniture: General Seating Solutions, L&B Industries
Carpets & Flooring: Constantine, Garden State Tile
Fabrics: ArcCom, Carnegie, Donghia, Majalite, Perennials, Silver State, Ultra Fabrics
Lighting: Architectural Details
Ceilings: Ralph Lauren Surfaces
Wallcoverings and Paint: Luxe Surfaces, Ralph Lauren Surfaces
General Contractors: G.B. Mannisto

Sands Macau Expansion, Casino
Design Firm: Steelman Partners
Furniture: World Sourcing Services Limited (Venetian Procurement)
Carpets & Flooring: Tai Ping Carpets
Fabrics: World Sourcing Services Limited (Venetian Procurement)
Lighting: Diamond Life, Lumitech
Ceilings: BSC
Wallcoverings and Paint: ICI, Original Specified Product
General Contractors: FiTout Contractor-BSC
Lighting Consultants: Paul Steelman Design Group

Seminole Hard Rock Hotel & Casino, Hollywood, FL
Design Firm: Klai Juba
Furniture: Blue Leaf Hospitality

Seminole Hard Rock Hotel & Casino, Tampa, FL
Design Firm: Klai Juba
Furniture: Blue Leaf Hospitality

Seneca Allegany Casino & Hotel
Design Firm: JCJ Architecture
Furniture: Artone, Charter Furniture, Donghia, Gasser Chair Co., ISA International, Lambert Furniture, Loewenstein Contract, Mark David, Sandler, Shafer, Thompson, Tropitone Contract, Vintage
Carpets & Flooring: Ulster Carpets
Fabrics & Surfaces: Anzea, ArcCom, Architex, Bisazza, Crossville Ceramics, Daltile, Designtex, Douglas Industries, Innovations in Textiles, Integra Fabrics, JM Lynne, Knoll, Textiles, Knox Tile, Kravet, Maharam, Panaz, Stone Source, Unika Vaev, Valley Forge Fabrics
Lighting: Alger, Hubbardton Forge, LightSpann, Morrison Lighting, Prima
Ceilings: Armstrong, Gage, USG
Wallcoverings and Paint: Archetonic, Benjamin Moore, Designtex, D.L. Couch, Innovations in Wall Covering, JM Lynne, Lanark, Mahaaram, MDC Wallcoverings, Pittsburgh Paint, Sanitas, Sherwin-Williams, Wolf Gordon

Seneca Niagara Casino and Hotel
Design Firm: JCJ Architecture
Furniture: Artone Manufacturing, Gasser, ISA International, Spa Equip, Shelby Williams, Thomasville Furniture, Vintage
Carpets & Flooring: Armstrong, Contiva, Crossville Ceramics, Daltile, Ulster Carpets
Fabrics: ArcCom, Architex, Liz Jordan-Hill, Maharam, Momentum
Wallcoverings and Paint: Benjamin Moore, Designtex, Innnnovations, JM Lynne, Koroseal, Lamark, Maharam, Seabrook, Sherwin-Williams, Wolf Gordon

Sheraton Hotel Bangalore
Design Firm: DiLeonardo International, Inc.
Furniture: B&B Italia, Brueton, HBF, Hightower, Matteo Grassi
Carpets & Flooring: Atlas, Constantine, Texstyle
Fabrics: Bergamo, Edelman Leather, Knoll, Kravet

Lighting: Berillo, Boyd
Wallcoverings and Paint: GKD Metal, Maya Romanoff, Wolf Gordon

Studio 54
Design Firm: Cleo Design
Furniture: Colber, Eric Brand, Industrial Interiors
Carpets & Flooring: Junkers Hardwood
Fabrics: Fabricut, Kirk Brummel, Krupnick Brothers
Lighting: FiberopticStudio
Window Treatments: Coast Drapery
General Contractors: Madison Building Group

Susanna Foo
Design Firm: Daroff Design
Furniture: Chairmasters, Jay Sanders, LB Furniture Industries, Table Topics
Carpets & Flooring: Catco Marble & Granite, Durkan
Fabrics: Architex/Liz Jordan Hill, Fabricut, Knoll Textiles, Meyer Contract, P. Kaufmann
Wallcoverings and Paint: Benjamin Moore
Window Treatments: Gold Brothers
General Contractors: Remcor
Lighting Consultants: Focus Lighting

Taneko Japanese Tavern
Design Firm: MBH Architects
General Contractors: Rick Story Construction

Tempo Lounge
Design Firm: Cleo Design
Furniture: La Char, Theming Solutions, 3-Form, Tucker Robins Ironworks, West Coast
Carpets & Flooring: Couristan, Maya Romanoff
Fabrics: Architex, Clarence House, Decor de Paris, Pindler and Pindler, Tiger Imports, Valley Forge, Zax, Zimmer Rhode
Lighting: Orion Chandeliers
Ceilings: Visual Impact Technology
Wallcoverings and Paint: Couristan, Maya Romanoff
Window Treatments: Coast Drapery
General Contractors: Vergith

Vaquero
Design Firm: BraytonHughes Design Studios
Furniture: Astoria, Century Furniture, Fong Brothers, International Ironworks, National Upholstery, Troy Wesnidge
Carpets & Flooring: European Limestone
Fabrics: Brunschwig & Fils, Clarence House, Cowtan + Tout, Donghia, Perennials
Lighting: Morrison Lighting, Orion, R. Jesse, Steven Handelman Studios
Wallcoverings and Paint: Alderwood Paneling
General Contractors: Parkway Construction
Lighting Consultants: Flack + Kurtz

Venetian Resort Casino, Venezia Tower
Design Firm: sfa design
Furniture: Kneedler Faucher & Gregorious Pineo
Carpets & Flooring: Verona-Miles of Marble
Fabrics: Scalamandre
Lighting: Alger
Wallcoverings and Paint: Designer's Art & Accessories, EverGreene Painting Studios

Westin Stamford, lobby, lounge and restaurant
Design Firm: Haverson Architecture and Design
Furniture: David Edwards, ISA International, Michael Gold, Texstyle
Carpets & Flooring: AKDO
Fabrics: ArcCom, Architex, Calvin Henry, Deepa, Momentum, Pollack, Symphony
Lighting: Crate and Barrel, Color Kinetics, Lightolier
Ceilings: Hibernian Millwork
Wallcoverings and Paint: Ralph Lauren
General Contractors: Haverson Construction Management LLC
Lighting Consultants: Haverson Architecture and Design

Wildfire
Design Firm: Aria Group Architects, Inc.
Furniture: E.J. Industries, ISA International, Shelby Williams, Wood Good
Carpets & Flooring: Brintons, U.S. Axminster
Fabrics: Fabricut
Lighting: Antoine Prouix, CL Sterling, Holly Hunt, Lake Shore Studios, New Metal Crafts
Ceilings: Armstrong, Huder Woodworking
Wallcoverings and Paint: Maya Romanoff, Wolf Gordon
Window Treatments: Knoll Textiles, Maharam
General Contractors: Crown Construction
Lighting Consultants: Schuler Shook

Wolfgang's Steakhouse
Design Firm: DAS Architects, Inc.
Carpets & Flooring: Mercer Wood Flooring
Fabrics: Majalite
Lighting: Architectural Details
Ceilings: Minos Ltd.
Wallcoverings and Paint: Benjamin Moore, Wolf Gordon
General Contractors: Barplex

Wynn Macau
Design Firm: HBA/Hirsch Bedner Associates
Furniture: Kingwood, Lewis Mittman
Carpets & Flooring: Thai Ping
Fabrics: Fabricut, Janet Yonaty, Osborne & Little
Lighting: Alger, Triton
Wallcoverings and Paint: ICI Paint, Innovations, Maya Romanoff, Novawall
Window Treatments: Turner Brothers
General Contractors: Leighton China State Joint Venture
Lighting Consultants: Desmond O'Donova, DHA Designs

Index by Project

The Designer Series

Visual Reference Publications, Inc.
302 Fifth Avenue, New York, NY 10001
212.279.7000 • Fax 212.279.7014
www.visualreference.com

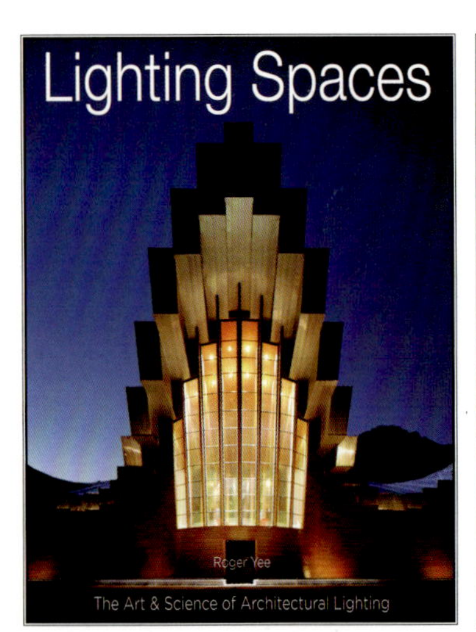

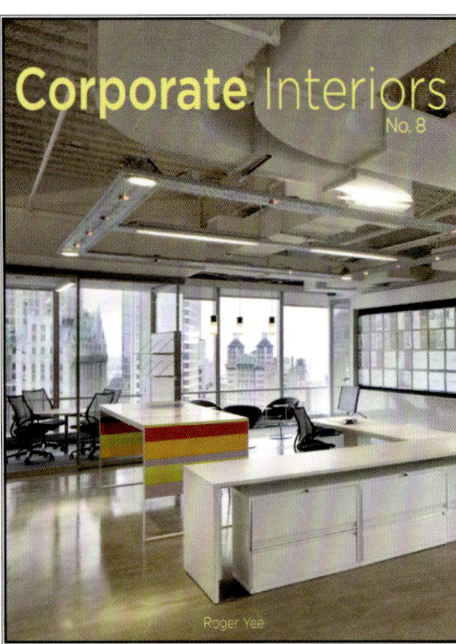

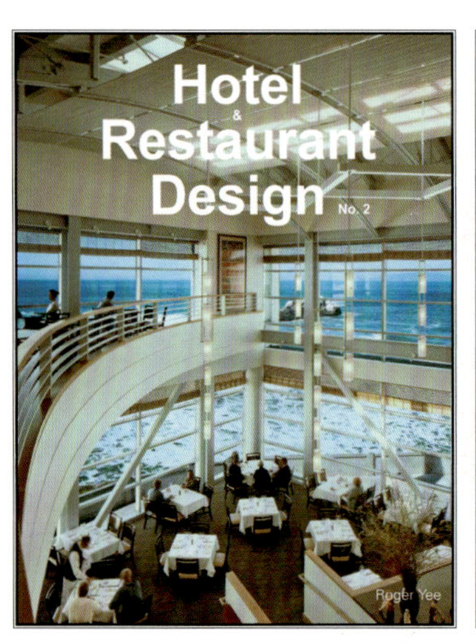

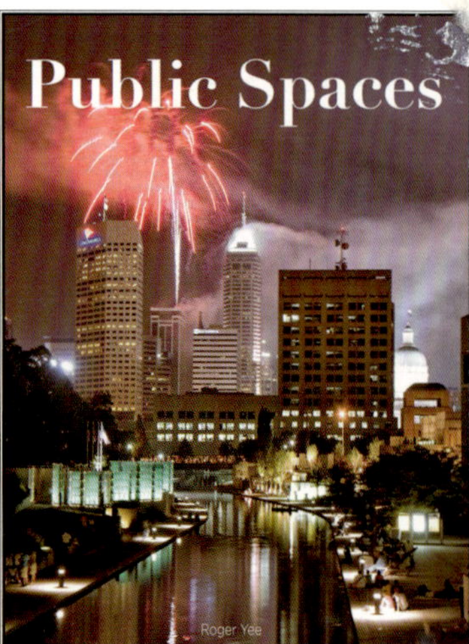

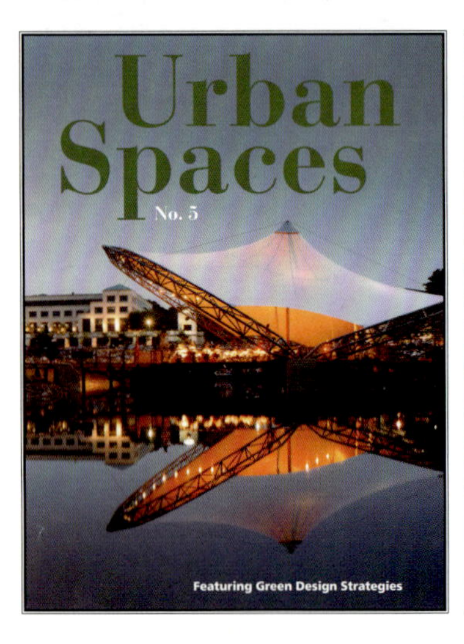

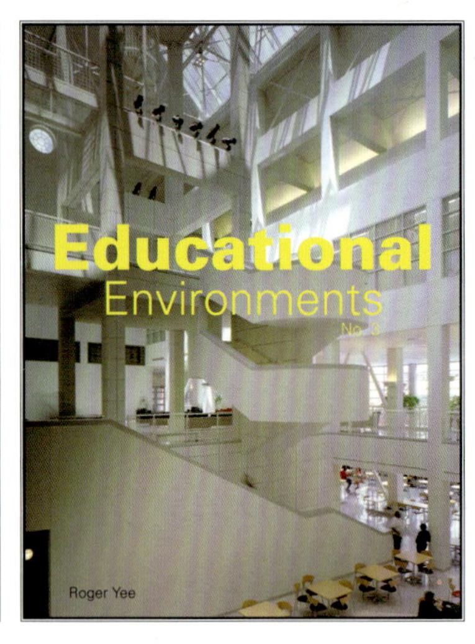

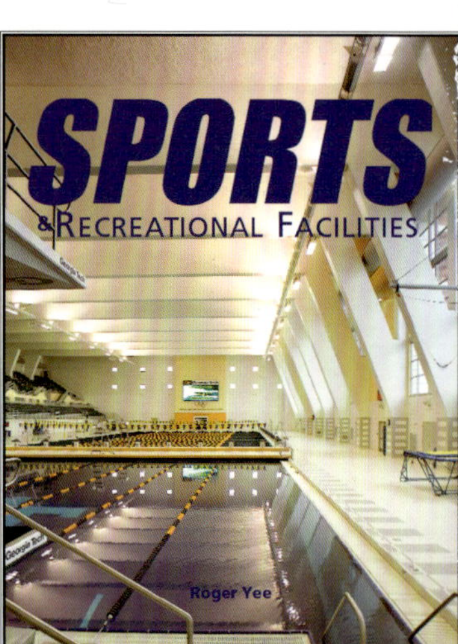